Land Surveyor-in-Training

Sample Examination

Second Edition

George M. Cole, PE, PLS

Professional Publications, Inc. • Belmont, CA

How to Get Online Updates for This Book

I wish I could claim that this book is 100% perfect, but 25 years of publishing have taught me that textbooks seldom are. Even if you only took one course in college, you are familiar with the issue of mistakes in textbooks.

I am inviting you to log on to Professional Publications' website at **www.ppi2pass.com** to obtain a current listing of known errata in this book. From the website home page, click on "Errata." Every significant known update to this book will be listed as fast as we can say "HTML." Suggestions from readers (such as yourself) will be added as they are received, so check in regularly.

PPI and I have gone to great lengths to ensure that we have brought you a high-quality book. Now, we want to provide you with high-quality after-publication support. Please visit us at **www.ppi2pass.com**.

Michael R. Lindeburg, PE
Publisher, Professional Publications, Inc.

LAND SURVEYOR-IN-TRAINING SAMPLE EXAMINATION
Second Edition

Printed in the United States of America

Professional Publications, Inc.
1250 Fifth Avenue, Belmont, CA 94002
(650) 593-9119
www.ppi2pass.com

Current printing of this edition: 2

Library of Congress Cataloging-in-Publication Data
Cole, George M.
    Land surveyor-in-training sample examination / George M. Cole.
        p.   cm.
    ISBN 1-888577-70-3
    1. Surveying--United States--Examinations--Study guides.   I. Title.
    TA537.C64   2001
    526.9′ 076--dc21
                                                                2001019565

TABLE OF CONTENTS

INTRODUCTION

DESCRIPTION OF THE EXAMINATION

In most of the United States, requirements for registration as a land surveyor include a two-day written examination. The first day is the Fundamentals of Land Surveying Examination, which is a standardized test prepared by the National Council of Examiners for Engineering and Surveying (NCEES). In many states, this test is a requirement for the Surveyor-In-Training or Surveying Intern registration.

The NCEES Fundamentals of Land Surveying Examination is an eight-hour examination concentrating on basic principles of land surveying and the application of mathematical formulas to the solution of surveying problems. The test consists of two four-hour sections, separated by a one-hour break. Each of the two portions of the examination contains 85 questions. For each question, you will be asked to select the best answer from four choices. Examination books and mechanical pencils are provided for recording answers, which are scored by machine. No credit is deducted for wrong answers. Therefore, it is in your best interest to answer each question.

The Fundamentals Examination is "closed book." No reference material of any kind may be used. However, each examination contains a collection of pertinent reference formulas that may be used. These include formulas for triangle solutions, horizontal and vertical curves, statistics, state plane coordinates, earthwork, tape correction, astronomy, photogrammetry, and stadia. In addition, some conversion factors are provided. Typical formulas that will be provided are included in this publication by permission of NCEES. However, it is emphasized that the formulas and conversion factors provided are not necessarily all-inclusive. Formulas and conversion factors other than those provided may be necessary to complete the examination. Therefore, a well-prepared examinee should know many of the basic formulas. Although rules may vary with individual state boards, calculators are usually allowed as long as they are battery-operated, silent, and nonprinting.

The Fundamentals Examination is not graded on a curve; a certain minimum competency must be demonstrated to safeguard the public welfare. Nevertheless, it is recognized that the tests may vary slightly in difficulty, depending upon the questions selected for a particular examination. Therefore, questions are reviewed by committees of practicing land surveyors before the examinations. These committees evaluate the difficulty of each question in order to develop a recommended passing score for each examination. However, the individual state boards have the authority to determine the passing score in their respective states. In the grading process, credit is given for each correct answer and no points are deducted for incorrect answers. The sum of the correct answers is scaled so that the grade of 70 reflects the standard minimum competency.

EXAMINATION CONTENTS

The Fundamentals Examination includes questions in 20 content areas, based on a recent task analysis of surveying practice. These areas are listed here, along with a list of typical surveying tasks that might be addressed in the Fundamentals Examination. Also provided is the percentage of examination questions that can be expected in each area. This information is based on recent NCEES publications.

Algebra & Trigonometry (6%)

This area includes units of measurement; formula development; formula manipulation; solving systems of equations; basic mensuration formulas for length, area, and volume; quadratic equations; trigonometric functions; right triangle solutions; oblique triangle solutions; trigonometric identities; spherical coordinates; and trigonometry.

Typical tasks in this area:

Perform trigonometric and differential leveling.
Compute survey data.
Compute areas and volumes.
Determine and prepare lot and street patterns for land division.
Design horizontal and vertical alignment for roads within a subdivision.

Higher Math (beyond trigonometry) (4%)

This area includes analytical geometry; linear algebra; equations of a line, circle, parabola, and ellipse; differentiation of functions; integration of elementary functions; infinite series; and mathematical modeling.

Typical tasks in this area:

Perform geodetic surveys using conventional methods and geodetic and/or plane surveys using GPS methods.
Perform astronomic measurements.
Compute, analyze and adjust survey data.
Design horizontal and vertical alignment for roads within a subdivision.

Probability & Statistics (4%)

This area includes standard deviation; variance; standard deviation of unit weight; tests of significance; concept of probability and confidence intervals; error ellipses; data distributions; and histograms.

Typical tasks in this area:

Determine levels of precision and order of accuracy.
Perform geodetic and/or plane surveying using GPS methods.
Compute, analyze, and adjust survey data.

Basic Sciences (3%)

This area includes light and wave propagation; basic electricity; optics; gravity; refraction; mechanics; forces; kinematics; temperature and heat; biology; dendrology; geology; and plant science.

Typical tasks in this area:

Calibrate instruments.

Geodesy & Survey Astronomy (4%)

This area includes reference ellipsoids; gravity fields; geoid; geodetic datums; direction and distance on the ellipsoid; conversion from geodetic heights to elevation; orbit determination and tracking; determination of azimuth using common celestial bodies; and time systems.

Typical tasks in this area:

Select appropriate vertical and/or horizontal datum and basis of bearing.
Perform geodetic surveys using conventional methods and geodetic and/or plane surveys using GPS methods.
Perform astronomic measurements.
Perform hydrographic surveys.
Perform differential leveling.
Analyze and adjust survey data.

Computer Operations & Programming (5%)

This area includes operating systems; graphical user interfaces (windows); data communication by serial or parallel interface; bits and bytes; Internet; computer architecture; keyboard programming of a handheld calculator; programming a computer in a compiled language; order of arithmetic operations; programming concepts such as decision statements; flow charts; looping; and arrays.

Typical tasks in this area:

Perform hydrographic surveys.
Produce survey data using photogrammetric methods.
Utilize survey data produced from photogrammetric methods.
Compute survey data.
Utilize computer-aided drafting systems.

Written & Verbal Communication (6%)

This area includes written communication; grammar; sentence structure; punctuation; bibliographical referencing; and verbal and nonverbal communication.

Typical tasks in this area:

Evaluate project elements to define scope of work.
Prepare and negotiate proposals and/or contracts.
Consult and coordinate with allied professionals and/or regulatory agencies.
Facilitate regulatory review and approval of project documents and maps.
Determine and secure entry rights.
Gather parol evidence.
Perform boundary surveys.
Advise clients regarding boundary uncertainties.
Testify as an expert witness.
Review documents with clients and/or attorneys.
Prepare sketches and/or preliminary plats, survey maps, plats, reports, and land descriptions.

Cadastral Law & Administration (6%)

This area includes land descriptions; real property rights; concepts of land ownership; case law; statute law; conveyancing; official records; public/quasi-public/private land record sources; tax assessment; legal instruments of title; the U.S. Public Land Survey System; colonial/metes and bounds survey system; subdivision survey system; and other cadastral systems.

Typical tasks in this area:

Facilitate regulatory review and approval of project documents and maps.
Determine and secure entry rights.
Research and evaluate court records and case law and evidence from private and public record sources.
Gather and evaluate parol evidence.
Perform boundary and condominium surveys.
Reconcile survey and record data.
Identify and evaluate field evidence for possession, boundary line discrepancies, and potential adverse possession claims.

Identify riparian and/or littoral boundaries.
Apply Public Land and other survey system principles.
Evaluate the priority of conflicting title elements.
Determine the location of boundary lines and encumbrances.
Advise clients regarding boundary uncertainties.
Testify as an expert witness.
Review documents with clients and/or attorneys.
Determine subdivision development requirements and constraints.
Determine and prepare lot and street patterns for land division.
Perpetuate and/or establish monuments and their records.
Document potential possession claims.
Prepare and file record of survey.

Boundary Law   (7%)

This area includes rules of evidence relative to land boundaries and court appearance; boundary control and legal principles; order of importance of conflicting title elements; possession principles; conflict resolution; riparian/littoral/water boundaries; boundary evidence; and simultaneous and sequential conveyance.

Typical tasks in this area:

Research and evaluate court records and case law and evidence from private and public record sources.
Gather and evaluate parol evidence.
Identify pertinent physical features, landmarks, and existing monumentation.
Perform boundary, route and right-of-way surveys.
Reconcile survey and record data.
Identify and evaluate field evidence for possession, boundary line discrepancies, and potential adverse possession claims.
Identify riparian and/or littoral boundaries.
Apply public land and other survey system principles.
Evaluate the priority of conflicting title elements.
Determine the location of boundary lines and encumbrances.
Advise clients regarding boundary uncertainties.
Testify as an expert witness.
Review documents with clients and/or attorneys.
Determine subdivision development requirements and constraints.
Perpetuate and/or establish monuments and their records.
Document potential possession claims.
Prepare survey maps, plats, reports, and land descriptions.

Business Law, Management, Economics, & Finance   (4%)

This area includes sole proprietorship, corporation, and partnership structures; contract law; tax structure; employment law; liability; operation analysis and optimization; land economics; appraisal science; critical path analysis; personnel management principles; cost/benefit analysis of a project or operation; econometric modeling; time value of money; and budgeting.

Typical tasks in this area:

Evaluate project elements to define scope of work.
Prepare and negotiate proposals and/or contracts.
Consult and coordinate with allied professionals and/or regulatory agencies.
Consult with and advise clients and/or their agents.
Facilitate regulatory review and approval of project documents and maps.

Surveying & Mapping History   (4%)

This area includes surveying/mapping instruments and their development; prominent events and personalities; history of cartography; photogrammetric instruments and their development; and history of the profession.

Typical tasks in this area:

Research and evaluate court records and case law and evidence from private and public record sources.
Gather parol evidence.
Recover horizontal/vertical control.
Calibrate instruments.
Apply Public Land and other survey system principles.
Evaluate the priority of conflicting title elements.
Determine locations of boundary lines and encumbrances.

Field Data Acquisition & Reduction   (6%)

This area includes field notes and electronic data collection; measurement of distances, angles and directions; modern instruments and their construction and use; precise levels; theodolites; total stations; EDMs; precision tapes; global positioning system; hydrographic data collection instruments; construction layout instruments and procedures for routes; and structures.

Typical tasks in this area:

Determine levels of precision and order of accuracy.
Recover horizontal/vertical control.
Identify pertinent physical features, landmarks, and existing monumentation.

Calibrate instruments.
Perform geodetic and plane surveys using conventional methods and geodetic and/or plane surveys using GPS methods.
Perform astronomic measurements.
Perform record, as-built, ALTA/ACSM, hydrographic, and photogrammetric control surveys.
Perform trigonometric or differential leveling.
Perform field verification of photogrammetric maps.
Produce survey data using photogrammetric methods.
Perform boundary, route and right-of-way, topographic, flood plain, construction, and condominium surveys.
Perpetuate and/or establish monuments and their records.

Photo/Image Data Acquisition & Reduction   (3%)

This area includes cameras; image scanners; digitizers; stereo plotters; orientation; editing; ortho-image production; geo-rectification; airborne GPS; image processing; and raster/vector data conversions.

Typical tasks in this area:

Determine levels of precision and orders of acuracy.
Perform record, as-built, ALTA/ACSM, photogrammetric control, topographic, and flood plain surveys.
Perform field verification of photogrammetric maps.
Produce survey data using photogrammetric methods.
Utilize survey data produced from photogrammetric methods.
Prepare survey maps, plats, and reports.

Graphical Communication, Mapping   (5%)

This area includes principles of effective graphical display of spatial information; preparation of sketches; scaled drawings; survey plats and maps; interpretation of features on three-dimensional drawings; principles of cartography and map projections; computer mapping; and use of overlays.

Typical tasks in this area:

Perform record, as-built, and ALTA/ACSM surveys.
Produce survey data using photogrammetric methods.
Utilize survey data produced from photogrammetric methods.
Prepare worksheets for analysis of surveys.
Utilize computer-aided drafting systems.

Determine and prepare lot and street patterns for land division.
Design horizontal and vertical alignment for roads within a subdivision.
Prepare sketches and/or preliminary plats.
Prepare and file record of survey.
Prepare survey maps, plats, and reports.
Develop and/or provide data for LIS/GIS.

Plane Survey Calculation   (7%)

This area includes computation and adjustment of traverses; COGO computation of boundaries; route alignments; construction; subdivision plats; and calculation of route curves and volumes.

Typical tasks in this area:

Determine levels of precision and orders of accuracy.
Calibrate instruments.
Perform geodetic and plane surveys using conventional methods and geodetic and/or plane surveys using GPS methods.
Perform astronomic measurements.
Perform record, as-built, ALTA/ACSM, hydrographic, photogrammetric control, boundary, route, right-of-way, topographic, flood plain, construction, and condominium surveys.
Perform trigonometric and differential leveling.
Compute, analyze, and adjust survey data.
Reconcile survey and record data.
Convert survey data to appropriate datum.
Prepare worksheets for analysis of surveys.
Determine locations of boundary lines and encumbrances.
Determine and prepare lot and street patterns for land division.
Design horizontal and vertical alignment for roads within a subdivision.
Develop and/or provide data for LIS/GIS.

Geodetic Survey Calculation   (5%)

This area includes calculation of position on a recognized coordinate system such as latitude/longitude; state plane coordinate systems and UTM; coordinate transformation; scale factors; and convergence.

Typical tasks in this area:

Select appropriate vertical and/or horizontal datum and basis of bearing.
Determine levels of precision and orders of accuracy.
Perform geodetic and plane surveys using conventional methods and geodetic and/or plane surveys using GPS methods.

Perform astronomic measurements.

Perform record, as-built, ALTA/ACSM, hydrographic, photogrammetric control, boundary, route, right-of-way, topographic, flood plain construction, and condominium surveys.

Perform trigonometric and differential leveling.

Produce survey data using photogrammetric methods.

Compute, analyze, and adjust survey data.

Convert survey data to appropriate datum.

Prepare worksheets for analysis of surveys.

Determine locations of boundary lines and encumbrances.

Develop and/or provide data for LIS/GIS.

Measurement Analysis & Data Adjustment (6%)

This area includes analysis of error sources; error propagation; control network analysis; blunder trapping and elimination; least squares adjustment; calculation of uncertainty of position; and accuracy standards.

Typical tasks in this area:

Determine levels of precision and orders of accuracy.

Perform geodetic and plane surveys using conventional methods and geodetic and/or plane surveys using GPS methods.

Perform astronomic measurements.

Perform record, as-built, ALTA/ACSM, hydrographic, photogrammetric control, boundary, route, right-of-way, topographic, flood plain, construction, and condominium surveys.

Perform trigonometric and differential leveling.

Compute, analyze, and adjust survey data.

Reconcile survey and record data.

Convert survey data to appropriate datum.

Prepare worksheets for analysis of surveys.

Determine locations of boundary lines and encumbrances.

Determine and prepare lot and street patterns for land division.

Design horizontal and vertical alignment for roads within a subdivision.

Develop and/or provide data for LIS/GIS.

Geographic Information System Concepts (4%)

This area includes spatial data storage and retrieval and analysis systems; relational database systems; spatial objects; attribute value measurement; data definitions; schemas; metadata concepts; coding standards; GIS analysis of polygons and networks; buffering; overlay; and spatial data accuracy standards.

Typical tasks in this area:

Utilize computer-aided drafting systems.

Perpetuate and/or establish monuments and their records.

Prepare and file records of surveys.

Prepare survey maps, plats, and reports.

Develop and/or provide data for LIS/GIS.

Land Development Principles (5%)

This area includes soil classifications and properties; hydrology and hydraulics; land planning and practices and laws controlling land use; drainage systems; construction methods; geometric and physical aspects of site analysis and design of land subdivisions; street alignment calculations; and application of subdivision standards to platting of land.

Survey Planning, Processes, & Procedures (6%)

This area includes techniques for planning and conducting surveys including boundary surveys; control surveys; hydrographic surveys; topographic surveys; route surveys; aerial surveys; construction surveys; and also issues related to professional liability, ethics and courtesy.

Typical tasks in this area:

Evaluate project elements to define scope of work.

Prepare and negotiate proposals and/or contracts.

Consult and coordinate with allied professionals and/or regulatory agencies.

Consult with and advise clients and/or their agents.

Facilitate regulatory review and approval of project documents and maps.

Determine and secure entry rights.

Advise clients regarding boundary uncertainties.

Testify as an expert witness.

Review documents with clients and/or attorneys.

Document potential possession claims.

Prepare survey maps, plats, and reports.

Develop and/or provide data for LIS/GIS.

PREPARATION FOR THE EXAMINATION

Preparation for the Fundamentals Examination should be considered a long-term project. As currently structured, the examination is both comprehensive and fast-paced. Rapid recall, discipline, stamina, and mastery of all areas to be covered are essential to succeed on the examination. Development of these skills may require months of preparation, in addition to the years of academic study and/or work experience necessary to qualify for the examination.

It is suggested that preparation follow these steps:

1. Review this publication to gain insight into the nature and content of the examination, as well as into typical questions that might be included.

2. Thoroughly review each of the major areas to be tested by carefully reading through various publications such as those listed in the following section, and by answering review questions and problems in such texts.

   It is a good idea to prepare a concise outline as you work through each area. In many areas, review courses are available that will be helpful during this stage of preparation. This review should be on a rigorous schedule to help you develop the discipline and stamina necessary to do well on the examination.

3. Take the sample examination contained within this publication to evaluate your readiness for the examination.

4. Work on any weak areas detected by your sample examination.

5. Conduct a final review of your notes.

SUGGESTED STUDY MATERIAL

Numerous texts are available that cover fundamentals of surveying. The following are several of the author's personal favorites, which offer comprehensive coverage of the areas to be tested on the examination. Edition numbers have been omitted since new editions are often issued. Areas for which these references are especially recommended are indicated in parentheses.

Brown, C.M., Robillard, W.G., and Wilson, D.A. *Boundary Control and Legal Principles.* New York: John Wiley & Sons, 1995.
(Legal principles)

Brown, C.M., Robillard, W.G., and Wilson, D.A. *Evidence and Procedures for Boundary Location.* New York: John Wiley & Sons, 2001.
(Legal principles, especially dealing with descriptions)

Buckner, R.B. *A Manual on Astronomic and Grid North.* Rancho Cordova, CA: Landmark Enterprises, 1984.
(Survey astronomy)

Bureau of Land Management. *Manual of Instruction for Surveys of Public Lands.* Washington, DC: Superintendent of Documents, 1973.
(Legal principles and techniques for public land surveys and retracements of such surveys)

Cole, G.M. *Water Boundaries.* New York, NY: John Wiley & Sons, 1996.
(Legal principles and computational techniques for riparian and littoral boundaries)

Colley, Barbara C. *Practical Manual of Land Development.* New York: McGraw-Hill, 1998.
(Land development)

Davis, R.E., Foote, F.S., Anderson, J.M., and Mikhail, E.M. *Surveying Theory and Practice.* New York: McGraw-Hill, 1981.
(Computations, measurements and research)

Denny, Milton E. *Surveyors and Engineers Small Business Handbook.* Tuscaloosa, AL: CED Technical Services, 1994.
(Management)

Harbin, A.L. *Land Surveyor Reference Manual.* Belmont, CA: Professional Publications, 2001.
(Computations, measurements, sample problems)

Hickerson, Thomas F. *Route Location and Design.* New York: McGraw-Hill, 1967.
(Route survey computations)

Lindeburg, Michael *Engineering Economic Analysis.* Belmont, CA: Professional Publications, 2001.
(Economics and finance)

Robillard, W.G. and Bouman, L.J. *Clark on Surveying and Boundaries.* Charlottesville, VA: The Michie Co., 1987.
(Legal principles)

Smith, James R. *Introduction to Geodesy.* New York: John Wiley & Sons, 1997.
(Geodesy)

Van Sickle, Jan *GPS for Land Surveyors.* Boca Raton, FL: Ann Arbor Press, 1996.
(GPS)

Wattles, W.C. *Land Survey Descriptions.* Orange, CA: Gurdon Wattles Publications, 1976.
(Land descriptions)

Wolf & Dewitt *Elements of Photogrammetry (with Applications in GIS).* New York: McGraw-Hill, 2000.
(Photogrammetry, GIS)

Wolf & Ghilani *Adjustment Computations.* New York: John Wiley & Sons, 1997.
(Adjustments)

ABOUT THE SAMPLE EXAMINATION

The following sample examination is designed to be representative of the NCEES Fundamentals of Land Surveying Examination in subject material, in level of difficulty, and in the type of questions. The examination is the same length as the NCEES examination. Therefore, it will be useful as a review for the examination, allow evaluation of one's level of preparedness, and provide familiarity with the examination process. If a candidate can comfortably pass the sample examination, he or she should be able to pass the actual examination. However, there is no guarantee that any problem in this publication has or ever will appear on any actual examination. Preparation should not be based solely on this sample examination, but should also include review of a wide range of reference material.

Solutions for each question are contained in the second portion of this publication. These will be useful in evaluating one's approach to the problems.

Note that at the beginning of each portion of the sample examination, there are four pages of reference formulas and factors. These pages are provided courtesy of NCEES and are said by that organization to be representative of those provided in the actual examination. However, formulas and conversion factors other than those provided may be necessary to complete the examination.

Throughout the examination, all azimuths are measured clockwise from north. Standard abbreviations used in the examination are as follows.

ac	acres	$^\circ$	degrees
ft	feet	$'$	minutes
in	inches	$''$	seconds
km	kilometers	Δ	difference
m	meters	ft^2	square feet
yd	yards	ft^3	cubic feet
ch	chain	yd^3	cubic yards
lk	link		
$\sum$	summation		

Name: _____

Last   First   Middle
                                    Initial

LAND SURVEYOR-IN-TRAINING SAMPLE EXAMINATION

MORNING SECTION

PART 1

Instructions

This is a "closed book" examination. Subject to rules established by your state, you may not use any reference material. You may use any approved, battery-powered, silent calculator. No writing tablets or loose papers are permitted. Sufficient room for scratch work is provided in this examination booklet. You are not permitted to share or exchange materials with other examinees.

This section of the exam consists of 85 problems, each worth one point. You will have four hours in which to complete this section. Your score will be determined by the number of problems you solve correctly. No points will be deducted for incorrect answers. It is to your advantage to answer every question. When permission has been given by your proctor, break the seal on the examination booklet and remove the answer sheet. Write your name immediately in the space indicated. Check that all pages are present and legible. If any part of this booklet is missing, your proctor will issue a new booklet.

All solutions must be entered on the answer sheet. No credit will be given for answers appearing in the examination booklet. Mark your answers with a No. 2 pencil. Do not use a pen. Marks must be dark and completely fill the "bubble." Record only one answer per problem; if you mark more than one answer, you will not receive credit for the problem. If you change an answer, be sure the old bubble is erased completely; incomplete erasures may be read as intended answers.

If you finish early, check your work and make sure you have correctly followed all instructions. After checking your answers, you may turn in your examination booklet and answer sheet and leave the examination room. Once you leave, you will not be permitted to return to work on your examination.

Do not work any problems from the afternoon section of the exam during the four hours of this exam.

WAIT FOR PERMISSION TO BEGIN.

LAND SURVEYOR-IN-TRAINING
SAMPLE EXAMINATION

MORNING SECTION

Name: _____

1 Ⓐ Ⓑ Ⓒ Ⓓ	18 Ⓐ Ⓑ Ⓒ Ⓓ	35 Ⓐ Ⓑ Ⓒ Ⓓ	52 Ⓐ Ⓑ Ⓒ Ⓓ	69 Ⓐ Ⓑ Ⓒ Ⓓ
2 Ⓐ Ⓑ Ⓒ Ⓓ	19 Ⓐ Ⓑ Ⓒ Ⓓ	36 Ⓐ Ⓑ Ⓒ Ⓓ	53 Ⓐ Ⓑ Ⓒ Ⓓ	70 Ⓐ Ⓑ Ⓒ Ⓓ
3 Ⓐ Ⓑ Ⓒ Ⓓ	20 Ⓐ Ⓑ Ⓒ Ⓓ	37 Ⓐ Ⓑ Ⓒ Ⓓ	54 Ⓐ Ⓑ Ⓒ Ⓓ	71 Ⓐ Ⓑ Ⓒ Ⓓ
4 Ⓐ Ⓑ Ⓒ Ⓓ	21 Ⓐ Ⓑ Ⓒ Ⓓ	38 Ⓐ Ⓑ Ⓒ Ⓓ	55 Ⓐ Ⓑ Ⓒ Ⓓ	72 Ⓐ Ⓑ Ⓒ Ⓓ
5 Ⓐ Ⓑ Ⓒ Ⓓ	22 Ⓐ Ⓑ Ⓒ Ⓓ	39 Ⓐ Ⓑ Ⓒ Ⓓ	56 Ⓐ Ⓑ Ⓒ Ⓓ	73 Ⓐ Ⓑ Ⓒ Ⓓ
6 Ⓐ Ⓑ Ⓒ Ⓓ	23 Ⓐ Ⓑ Ⓒ Ⓓ	40 Ⓐ Ⓑ Ⓒ Ⓓ	57 Ⓐ Ⓑ Ⓒ Ⓓ	74 Ⓐ Ⓑ Ⓒ Ⓓ
7 Ⓐ Ⓑ Ⓒ Ⓓ	24 Ⓐ Ⓑ Ⓒ Ⓓ	41 Ⓐ Ⓑ Ⓒ Ⓓ	58 Ⓐ Ⓑ Ⓒ Ⓓ	75 Ⓐ Ⓑ Ⓒ Ⓓ
8 Ⓐ Ⓑ Ⓒ Ⓓ	25 Ⓐ Ⓑ Ⓒ Ⓓ	42 Ⓐ Ⓑ Ⓒ Ⓓ	59 Ⓐ Ⓑ Ⓒ Ⓓ	76 Ⓐ Ⓑ Ⓒ Ⓓ
9 Ⓐ Ⓑ Ⓒ Ⓓ	26 Ⓐ Ⓑ Ⓒ Ⓓ	43 Ⓐ Ⓑ Ⓒ Ⓓ	60 Ⓐ Ⓑ Ⓒ Ⓓ	77 Ⓐ Ⓑ Ⓒ Ⓓ
10 Ⓐ Ⓑ Ⓒ Ⓓ	27 Ⓐ Ⓑ Ⓒ Ⓓ	44 Ⓐ Ⓑ Ⓒ Ⓓ	61 Ⓐ Ⓑ Ⓒ Ⓓ	78 Ⓐ Ⓑ Ⓒ Ⓓ
11 Ⓐ Ⓑ Ⓒ Ⓓ	28 Ⓐ Ⓑ Ⓒ Ⓓ	45 Ⓐ Ⓑ Ⓒ Ⓓ	62 Ⓐ Ⓑ Ⓒ Ⓓ	79 Ⓐ Ⓑ Ⓒ Ⓓ
12 Ⓐ Ⓑ Ⓒ Ⓓ	29 Ⓐ Ⓑ Ⓒ Ⓓ	46 Ⓐ Ⓑ Ⓒ Ⓓ	63 Ⓐ Ⓑ Ⓒ Ⓓ	80 Ⓐ Ⓑ Ⓒ Ⓓ
13 Ⓐ Ⓑ Ⓒ Ⓓ	30 Ⓐ Ⓑ Ⓒ Ⓓ	47 Ⓐ Ⓑ Ⓒ Ⓓ	64 Ⓐ Ⓑ Ⓒ Ⓓ	81 Ⓐ Ⓑ Ⓒ Ⓓ
14 Ⓐ Ⓑ Ⓒ Ⓓ	31 Ⓐ Ⓑ Ⓒ Ⓓ	48 Ⓐ Ⓑ Ⓒ Ⓓ	65 Ⓐ Ⓑ Ⓒ Ⓓ	82 Ⓐ Ⓑ Ⓒ Ⓓ
15 Ⓐ Ⓑ Ⓒ Ⓓ	32 Ⓐ Ⓑ Ⓒ Ⓓ	49 Ⓐ Ⓑ Ⓒ Ⓓ	66 Ⓐ Ⓑ Ⓒ Ⓓ	83 Ⓐ Ⓑ Ⓒ Ⓓ
16 Ⓐ Ⓑ Ⓒ Ⓓ	33 Ⓐ Ⓑ Ⓒ Ⓓ	50 Ⓐ Ⓑ Ⓒ Ⓓ	67 Ⓐ Ⓑ Ⓒ Ⓓ	84 Ⓐ Ⓑ Ⓒ Ⓓ
17 Ⓐ Ⓑ Ⓒ Ⓓ	34 Ⓐ Ⓑ Ⓒ Ⓓ	51 Ⓐ Ⓑ Ⓒ Ⓓ	68 Ⓐ Ⓑ Ⓒ Ⓓ	85 Ⓐ Ⓑ Ⓒ Ⓓ

TRIGONOMETRIC FORMULAS

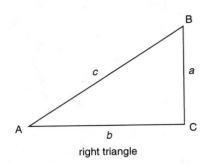

right triangle

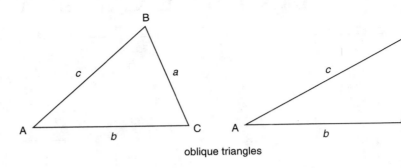

oblique triangles

Solution of Right Triangles

For angle A: $\sin A = \dfrac{a}{c}$ $\cos A = \dfrac{b}{c}$ $\tan A = \dfrac{a}{b}$ $\cot A = \dfrac{b}{a}$ $\sec A = \dfrac{c}{b}$ $\operatorname{cosec} A = \dfrac{c}{a}$

Given	Required		
a, b	A, B, c	$\tan A = \dfrac{a}{b} = \cot B$	$c = \sqrt{a^2 + b^2} = a\sqrt{1 + \dfrac{b^2}{a^2}}$
a, c	A, B, b	$\sin A = \dfrac{a}{c} = \cos B$	$b = \sqrt{(c + a)(c - a)} = c\sqrt{1 - \dfrac{a^2}{c^2}}$
A, a	B, b, c	$B = 90° - A$	$b = a \cot A$ $c = \dfrac{a}{\sin A}$
A, b	B, a, c	$B = 90° - A$	$a = b \tan A$ $c = \dfrac{b}{\cos A}$
A, c	B, a, b	$B = 90° - A$	$a = c \sin A$ $b = c \cos A$

Solution of Oblique Triangles

Given	Required		
A, B, a	b, c, C	$b = \dfrac{a \sin B}{\sin A}$	$C = 180° - (A + B)$ $c = \dfrac{a \sin C}{\sin A}$
A, a, b	B, c, C	$\sin B = \dfrac{b \sin A}{a}$	$C = 180° - (A + B)$ $c = \dfrac{a \sin C}{\sin A}$
a, b, C	A, B, c	$A + B = 180° - C$	$\tan\left(\dfrac{A - B}{2}\right) = \left(\dfrac{a - b}{a + b}\right)\left[\tan\left(\dfrac{A + B}{2}\right)\right]$
			$c = \dfrac{a \sin C}{\sin A}$
a, b, c	A, B, C	$s = \dfrac{a + b + c}{2}$	$\sin\left(\dfrac{A}{2}\right) = \sqrt{\dfrac{(s - b)(s - c)}{bc}}$
		$\sin\left(\dfrac{B}{2}\right) = \sqrt{\dfrac{(s - a)(s - c)}{ac}}$	$C = 180° - (A + B)$
a, b, c	area	$s = \dfrac{a + b + c}{2}$	$\text{area} = \sqrt{s(s - a)(s - b)(s - c)}$
A, b, c	area	$\text{area} = \dfrac{bc \sin A}{2}$	
A, B, C, a	area	$\text{area} = \dfrac{a^2 \sin B \sin C}{2 \sin A}$	

HORIZONTAL CURVE FORMULAS

D = degree of curve, arc definition

$1°$ = 1-degree of curve

$2°$ = 2-degree of curve

PC = point of curve

PT = point of tangent

PI = point of intersection

I = intersection of angle;
    angle between two tangents

L = length of curve
    from PC to PT

T = tangent distance

E = external distance

R = radius

LC = length of long chord

M = length of middle ordinate

c = length of subchord

d = angle of subchord

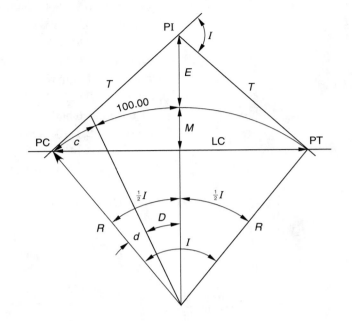

$$R = \frac{LC}{2\sin\left(\frac{I}{2}\right)} \qquad T = R\tan\left(\frac{I}{2}\right) = \frac{LC}{2\cos\left(\frac{I}{2}\right)}$$

$$\frac{LC}{2} = R\sin\left(\frac{I}{2}\right) \qquad D\,1° = R = 5729.58 \qquad D\,2° = \frac{5729.58}{2} \qquad D = \frac{5729.58}{R}$$

$$M = R\left[1 - \cos\left(\frac{I}{2}\right)\right] = R - R\cos\left(\frac{I}{2}\right)$$

$$\frac{E+R}{R} = \sec\left(\frac{I}{2}\right) \qquad \frac{R-M}{R} = \cos\left(\frac{I}{2}\right)$$

$$c = 2R\sin\left(\frac{d}{2}\right) \qquad d = \frac{c}{2R}$$

$$LC = 2R\sin\left(\frac{I}{2}\right) \qquad E = R\left[\sec\left(\frac{I}{2}\right) - 1\right] = R\sec\left(\frac{I}{2}\right) - R$$

MISCELLANEOUS CONVERSION FACTORS

1 meter = 3.280833 survey feet

1 survey foot = 0.3048006 meters

1 link = 0.66 feet

80 chains = 320 rods = 5280 feet

1 acre = 43,560 square feet

640 acres = 1 square mile

1 kilometer = 0.621370 mile

1 mile = 1.60935 kilometers

1 hectare = 2.47104 acres

π = 3.14159

1 radian = 57.2958 degrees

VERTICAL CURVE FORMULAS

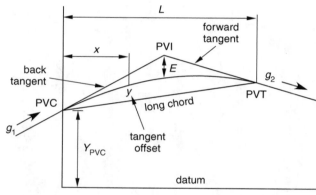

not to scale

L = length of curve

PVC = point of vertical curvature

PVI = point of vertical intersection

PVT = point of vertical tangency

g_1 = grade of back tangent

g_2 = grade of forward tangent

a = parabola constant

y = tangent offset

E = tangent offset at PVI

r = rate of change of grade

Y_{PVC} = elevation of PVC

$$y = ax^2 \qquad\qquad a = \frac{g_2 - g_1}{2L}$$

$$E = a\left(\frac{L}{2}\right)^2 \qquad\qquad r = \frac{g_2 - g_1}{L}$$

tangent elevation = $Y_{PVC} + g_1 x$

grade elvation = $Y_{PVC} + g_1 x + ax^2$

REFERENCE FORMULAS

Earthwork Formulas

$$\text{volume} = \frac{L(A_1 + A_2)}{2} \quad \text{[average end area formula]}$$

$$\text{volume} = \frac{L(A_1 + 4A_m + A_z)}{6} \quad \text{[prismoidal formula]}$$

$$\text{volume} = \frac{h(\text{area of base})}{3} \quad \text{[pyramid or cone]}$$

Area Formulas

$$\text{area} = \frac{X_A(Y_B - Y_N) + X_B(Y_C - Y_A) + X_C(Y_D - Y_B) + \cdots + X_N(Y_1 - Y_{N-1})}{2} \quad \text{[area by coordinates]}$$

$$\text{area} = w\left[\frac{h_1 + h_n}{2} + h_2 + h_3 + h_4 + \cdots + h_{n-1}\right] \quad \text{[trapezoidal rule]}$$

$$\text{area} = \frac{w[h_1 + 2(\sum h_{\text{odds}}) + 4(\sum h_{\text{evens}}) + h_n]}{3} \quad \text{[Simpson's one-third rule]}$$

Quadratic Equation Formula

$$ax^2 + bx + c = 0$$

$$x = \frac{-b \pm \sqrt{b^2 - 4ac}}{2a}$$

Correction Formulas

$$\text{correction for temperature} = C_t = 6.5 \times 10^{-6}(T - T_s)L$$

$$\text{correction for tension} = C_p = \frac{(P - P_s)L}{AE}$$

$$\text{correction for sag} = C_s = \frac{w^2 l^3}{24P^2}$$

T = temperature of tape during measurement, °F
T_s = temperature of tape during calibration, °F
L = distance measured, in ft
P = pull applied during measurement, in lb
P_s = pull applied during calibration, in lb
A = cross-section area of tape, in in^2
E = modulus of elasticity of tape, in psi
w = weight of tape, in lb per ft
l = length of unsupported span, in ft

Astronomy Formulas

$$\cos(\text{azimuth of sun}) = \frac{\sin\delta - \sin\phi\sin h}{\cos\phi\cos h}$$

$$Z = \text{bearing of Polaris at elongation} = \frac{p}{\cos\phi}$$

δ = declination angle

ϕ = latitude angle

h = altitude angle

p = polar distance of Polaris, angle

Photogrammetry Formulas

$$\text{average scale} = \frac{H - h_{\text{avg}}}{f}, \text{ in ft per in}$$

f = focal length of camera lens, in in

$H - h_{\text{avg}}$ = flying height above average ground surface, in ft

Stadia Formulas

$$\text{horizontal distance} = KS\cos^2\alpha$$

$$\text{vertical distance} = KS\sin\alpha\cos\alpha$$

K = stadia constant

S = rod intercept

α = vertical angle of sight

1. On a topographic map with a scale of 1:100,000, a distance of 14,000 ft on the ground would be represented by what distance on the map?

    (A) 0.5 in

    (B) 1.68 in

    (C) 2.5 cm

    (D) 7.14 in

2. The azimuth from a point to a reference mark is $264°26'50''$ as measured from astronomic north. If the Laplace correction is disregarded and the convergence is $+0°26'42''$, what is the state plane coordinate grid azimuth to the reference mark?

    (A) $264°00'08''$

    (B) $264°26'50''$

    (C) $264°53'32''$

    (D) due north

3. When fast land is formed by the gradual, natural lowering of the water level, what is the process called?

    (A) accretion

    (B) avulsion

    (C) reliction

    (D) riparian rights

4. For a five-sided, closed traverse with an angular misclosure of 10 sec, what is the sum of the interior angles?

    (A) $360°00'10''$ or $359°59'50''$

    (B) $480°00'10''$ or $479°59'50''$

    (C) $540°00'10''$ or $539°59'50''$

    (D) $720°00'10''$ or $719°59'50''$

5. If the bearing of a tangent entering a highway curve to the right is N45°E and the interior angle of the curve is 15°, what is the bearing of the tangent out of the curve?

    (A) N30°E

    (B) S45°E

    (C) N75°E

    (D) N60°E

6. A distance of 350.25 m is equivalent to which of the following?

    (A) 16.9 ch

    (B) 1050.8 ft

    (C) 1149.1 ft

    (D) 1150.8 ft

7. What is the width of a rectangle with a length of 400 ft and an area of 1 ac?

    (A) 108.7 ft

    (B) 108.9 ft

    (C) 109.3 ft

    (D) 110.2 ft

8. Which of the following is the earliest land information system?

    (A) Napoleonic Cadastre in France

    (B) Torrens System in Australia

    (C) Domesday Book in England

    (D) Grundbuch in Germany

9. A distance of 30 chains is equivalent to which of the following?

    (A) 120 rods

    (B) 604 m

    (C) 1980 ft

    (D) both (A) and (C)

10. The hour angle method for determining an azimuth by solar observation requires which of the following measurements?

    (A) time and vertical angle of the sun

    (B) time, and horizontal and vertical angle of the sun

    (C) horizontal and vertical angle of the sun

    (D) time and horizontal angle of the sun

11. The declination of a celestial body is analogous to which of the following?

    (A) latitude

    (B) right ascension

    (C) longitude

    (D) local hour angle

12. In a right triangle with two sides (other than the hypotenuse) of 100 ft and 200 ft, what is the smallest angle?

    (A) $26°33'54''$

    (B) $30°25'52''$

    (C) $45°13'24''$

    (D) $63°26'06''$

13. Considering the series of measurements 29.25, 29.23, 29.25, 29.27, 29.21, and 29.25, the value 29.24 represents which of the following?

(A) standard deviation

(B) median

(C) mean

(D) mode

14. Which of the following characteristics is unique to a patent?

(A) a conveyance of any interest the grantor has

(B) a conveyance guaranteeing lawful ownership

(C) an unwritten transfer of land

(D) a conveyance from a sovereign power

15. What is the bearing and distance (in the same units as the coordinates) between Point 1 and Point 2, having the following Cartesian coordinates?

Point	x-Coordinate	y-Coordinate
1	100.00	100.00
2	300.75	−150.25

(A) S21°51′E, 300.75

(B) S38°44′E, 320.82

(C) N51°16′W, 300.75

(D) S51°16′E, 320.82

16. When upland is gradually worn away by the action of moving water, what is the process called?

(A) accretion

(B) avulsion

(C) erosion

(D) reliction

17. In most states, which of the following is recognized as the boundary between submerged land under navigable tidal waters and bordering uplands?

(A) mean higher high water

(B) mean high water

(C) mean lower low water

(D) mean low water

18. What is the period used to determine average tidal heights in the U.S.?

(A) 28 days

(B) 1 year

(C) 16 years

(D) 19 years

19. If the horizontal distance between two targets is 1000 ft, what is the distance between the targets on a rectified aerial that has been enlarged four times from a negative with a scale of 1 in = 400 ft?

(A) 1.6 in

(B) 2.5 in

(C) 5 in

(D) 10 in

20. An asset is purchased for $25,000. It has an estimated economic life of five years, after which it will be sold for $5000. What is the depreciation, using the straight-line method, during the third year?

(A) $3000

(B) $4000

(C) $4500

(D) $5000

21. A level with a height of instrument of 10.000 ft is used to site a rod, which is being held on a benchmark 10° off plumb. The rod reading is 10.000 ft. What is the elevation of the benchmark?

(A) 0.000 ft

(B) 0.152 ft

(C) 8.264 ft

(D) 9.848 ft

22. In order to apply state plane coordinates, a distance measured at an elevation of 4000 ft has been reduced to sea level. If the distance is 2.5 mi when reduced to sea level and the earth has a radius of 20,906,000 ft, what is the distance at 4000 ft?

(A) 13,197.47 ft

(B) 13,200.00 ft

(C) 13,202.53 ft

(D) 13,204.12 ft

23. If the area of a forest as measured from an aerial photograph with a scale of 1 in = 400 ft is 20 in^2, what is the area of the forest?

(A) 3.7 ac

(B) 36.7 ac

(C) 62.4 ac

(D) 73.5 ac

24. The bearing of the tangent into a highway curve with a radius of 1000 ft is N60°E and the coordinates, in feet, of the PC are N = 1000, E = 1000. What are the coordinates, in feet, for the radius point?

    (A) N = 134.0, E = 1500.0

    (B) N = 144.0, E = 1500.0

    (C) N = 1500.0, E = 134.0

    (D) N = 1500.0, E = 144.0

25. Which expression best describes the relationship between the lengths of a second of arc of latitude and a second of arc of longitude in the U.S.?

    (A) latitude > longitude

    (B) longitude > latitude

    (C) latitude = longitude

    (D) (A), (B), or (C), depending on location

26. In a state plane coordinate system, grid north lies in which direction from true north for a point that is east of the central meridian of the system?

    (A) east

    (B) west

    (C) grid north is identical to true north

    (D) (A) or (B), depending on latitude

27. A platted subdivision consists of a series of 100-ft by 100-ft square lots except for the corner lots, which differ only in that one corner of the lots is a curve with a 30-ft radius and a central angle of 90°. What is the area of the corner lots?

    (A) 9100 ft^2

    (B) 9147 ft^2

    (C) 9807 ft^2

    (D) 10,000 ft^2

28. Which is the controlling call in the following description?

> ... S88°E a distance of 412 ft to the shore of Cripple Creek...

    (A) S88°E

    (B) 412 ft

    (C) the thread of Cripple Creek

    (D) the shore of Cripple Creek

29. The plat of an original 1950 government subdivision of Township 1 North, Range 2 East shows the south line of Section 30 to be 79.60 chains. What should be the length of the south line of the SW$\frac{1}{4}$ of the SW$\frac{1}{4}$ (Lot 4) of Section 30?

    (A) 10.00 ch

    (B) 20.00 ch

    (C) 19.60 ch

    (D) 39.50 ch

30. For the township described in Question 29, what should be the length of the south line of the SE$\frac{1}{4}$ of the SW$\frac{1}{4}$ of Section 30?

    (A) 10.00 ch

    (B) 19.60 ch

    (C) 20.00 ch

    (D) 39.50 ch

31. For the township described in Question 29, original monuments have been found at both the southwest and southeast corners of Section 29. The distance between the monuments is found to be 5278.20 ft. The quarter corner between the two found monuments is a lost corner, and no other monuments have been established on that line. What should be the length of the south line of the SW$\frac{1}{4}$ of the SW$\frac{1}{4}$ of Section 29?

    (A) 1319.55 ft

    (B) 1320.00 ft

    (C) 2639.10 ft

    (D) 2640.00 ft

32. What is the area of the illustrated polygon?

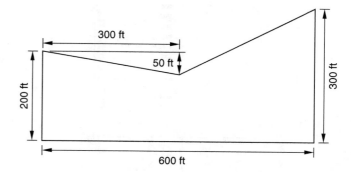

    (A) 110,000 ft^2

    (B) 115,500 ft^2

    (C) 120,000 ft^2

    (D) 142,500 ft^2

33. What is a major advantage of dual-frequency over single-frequency GPS receivers?

    (A) precise measurement of longer baselines

    (B) observation of more satellites

    (C) observation in bad weather conditions

    (D) measurement in three dimensions

34. What method should be used to restore a lost southeast corner of Section 31 in a typical township?

    (A) bearing-bearing intersection

    (B) single proportional measurement

    (C) double proportional measurement

    (D) record bearing and distance

35. The star Polaris is located in which constellation?

    (A) Cassiopeia

    (B) Ursa Minor

    (C) Ursa Major

    (D) Orion

36. What is the length of one link of a Gunter's chain?

    (A) 7.9 in

    (B) 0.3 m

    (C) 1 ft

    (D) 66 ft

37. A level rod reads as shown. Assuming a stadia constant of 100, what is the approximate distance to the rod, in the units of the rod?

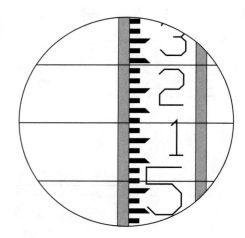

    (A) 23.4 ft

    (B) 51.3 ft

    (C) 117.0 ft

    (D) 234.0 ft

38. When sighting through a leveled instrument at a distant level rod, curvature of the earth would have what effect on the rod reading?

    (A) There would be no effect.

    (B) The rod reading would be greater.

    (C) The rod reading would be less.

    (D) The effect would vary with stadia constant.

39. What would be the effect of refraction on the rod reading in Question 38?

    (A) There would be no effect.

    (B) The rod reading would be greater.

    (C) The rod reading would be less.

    (D) The effect would vary with stadia constant.

40. On a map developed from a Lambert projection, which of the following is true?

    (A) Meridians converge.

    (B) Parallels are arcs of circles.

    (C) Parallels are uniformly spaced.

    (D) (A), (B), and (C) are true.

41. Which method should be used to restore a lost southwest corner of Section 31 in a typical township?

    (A) bearing-bearing intersection

    (B) single proportional measurement

    (C) double proportional measurement

    (D) record bearing and distance

42. Which of the following can be accomplished by the DMD method?

    (A) balancing a closed traverse

    (B) correcting latitude observation

    (C) correcting longitude observation

    (D) determining area of a closed figure

43. The following illustration represents a closed traverse with the only significant error in measurement being one angular error. At which station was the error most likely made?

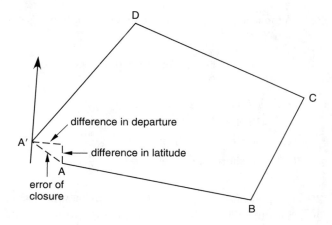

(A) A

(B) B

(C) C

(D) D

44. A closed traverse has a total length of 2181.14 ft and an error of closure of 0.37 ft in the northing and 0.24 ft in the easting. What is the ratio of error for the closure?

(A) 1:4957

(B) 1:5187

(C) 1:6319

(D) 1:9453

45. Elevations provided for geodetic benchmarks are usually referenced to which datum?

(A) mean tide level

(B) mean sea level

(C) National Geodetic Vertical Datum

(D) North American Datum of 1927

46. What is the arc length of a highway curve with a radius of 954.93 ft, an interior angle of 60°, and the following stationing?

PC	23 + 45.67
PI	28 + 97.00
PT	33 + 45.67

(A) 551.33 ft

(B) 954.93 ft

(C) 957.82 ft

(D) 1000.00 ft

47. A grid distance in a state plane coordinate system is 5280.00 ft and the scale factor is 0.9999424. What is the surface distance, assuming no sea level correction is necessary?

(A) 5278.64 ft

(B) 5279.70 ft

(C) 5280.00 ft

(D) 5280.30 ft

48. A compound highway curve, with both curves having interior angles of 45°, has an approach tangent of N45°E. What is the bearing of the tangent leaving the curves?

(A) N45°E

(B) S45°E

(C) north

(D) east

49. On the topographic map illustrated, what is the highest elevation?

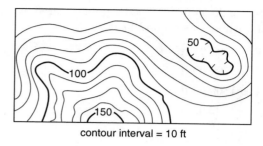

contour interval = 10 ft

(A) 148 ft

(B) 150 ft

(C) 152 ft

(D) 161 ft

50. On the topographic map illustrated in Question 49, what is the lowest elevation?

(A) 43 ft

(B) 50 ft

(C) 54 ft

(D) 61 ft

51. Observed from a height of instrument of 1000.00 ft with a zenith angle of 87°, a distant target is observed at a horizontal distance of 5280 ft. If the effects of curvature and refraction are not considered, what is the approximate elevation of the target?

(A) 723.29 ft

(B) 1276.71 ft

(C) 1524.92 ft

(D) 5372.76 ft

52. Which of the following describes a method of obtaining easement rights from long-continued usage?

(A) adverse possession

(B) eminent domain

(C) prescription

(D) dedication

53. Which of the following describes a method of acquiring title to property by long-term occupation?

(A) adverse possession

(B) eminent domain

(C) prescription

(D) dedication

54. Assuming a straight segment of highway centerline, what is the horizontal distance between a point on the right-of-way of a highway located at station 32+42 with an offset of 20.55 ft right and a point located at station 50+41 with an offset of 42.12 ft left?

(A) 1797.91 ft
(B) 1799.00 ft
(C) 1800.09 ft
(D) 1802.91 ft

55. Given the following three-wire leveling readings (in feet) and backsight elevation of 100.00 ft, what is the elevation of the foresight?

BS	FS
5.242	6.813
4.495	6.060
3.751	5.302

(A) 98.438 ft
(B) 98.455 ft
(C) 101.562 ft
(D) 101.565 ft

56. Given the readings in Question 55, what is the approximate backsight distance?

(A) 75 ft
(B) 76 ft
(C) 149 ft
(D) 151 ft

57. Which of the following is the standard width of a time zone in degrees of longitude?

(A) 7.5°
(B) 10°
(C) 12°
(D) 15°

58. The bearing from a backsight to an instrument station is S55°E. An angle of 75° is turned to the right. What is the bearing from the station to the foresight?

(A) S20°W
(B) N20°E
(C) S40°W
(D) N40°E

59. A vertical highway curve is usually a portion of which type of curve?

(A) parabola
(B) circle
(C) hyperbola
(D) spiral

60. On a map with a plane projection with the plane parallel to the equator and tangent to the north pole, which of the following is true?

(A) Meridians converge at the pole.
(B) Parallels are straight lines.
(C) The distance between parallels increases with latitude.
(D) All great circles may be plotted as straight lines.

61. An aerial photograph is taken from an altitude of 2400 ft, using a camera with a focal length of 6 in. What is the scale of the negative?

(A) 1 in = 400 ft
(B) 1 in = 500 ft
(C) 1 in = 1200 ft
(D) 1 in = 4800 ft

62. The recorded plat of the Live Oak Subdivision shows Lot 10, Block B to be a 100-ft by 200-ft rectangular lot. The original owner of the lot conveyed the west one-half of the lot in 1960 and the eastern 100 ft of the lot in 1990. Neither conveyance has been previously surveyed. A current survey of the 1990 conveyance finds that the north and south lines of Lot 10 measure 198.5 ft. Where should the northwest corner of the land conveyed in 1990 be placed?

(A) 89.25 ft from the northeast corner
(B) 98.50 ft from the northeast corner
(C) 99.25 ft from the northwest corner
(D) 100 ft from the northwest corner

63. Why is the answer for Question 62 correct?

(A) Fractional conveyances control over distances.
(B) Senior rights control.
(C) Most-current conveyance controls.
(D) Recorded plat distances control over measured distances.

64. If, in Question 62, both the north and south lines of Lot 10 were found to be 202.5 ft in length, where should the northwest corner of the land conveyed in 1990 be placed?

(A) 99.25 ft from the northeast corner

(B) 100 ft from the northeast corner

(C) 100 ft from the northwest corner

(D) 101.25 ft from the northeast corner

65. If the Federal Geodetic Control Commission standards for level loop closures are as follows, a state standard requiring a closure of ± 0.05 ft $\times \sqrt{\text{miles}}$ is equivalent to which order of accuracy?

Order	Closure
First Order, Class I	$(4\text{ mm})(\sqrt{K})$
First Order, Class II	$(5\text{ mm})(\sqrt{K})$
Second Order, Class I	$(6\text{ mm})(\sqrt{K})$
Second Order, Class II	$(8\text{ mm})(\sqrt{K})$
Third Order	$(12\text{ mm})(\sqrt{K})$

$(K = \text{distance in km})$

(A) First Order, Class II

(B) Second Order, Class I

(C) Second Order, Class II

(D) Third Order

66. In a retracement survey of government Section 12, neither the original monument nor accessories for the northeast corner can be found. An iron pipe is found at the corner, which the survey finds is in harmony with adjacent corners. An adjacent landowner indicates that he set the iron pipe at the former location of the original wood monument. Which of the following describes the original corner?

(A) existent

(B) lost

(C) obliterated

(D) reset by double proportion

67. Why is the answer to Question 66 correct?

(A) The landowner's testimony validates the location of the iron pipe.

(B) The iron pipe cannot be accepted unless set by a surveyor.

(C) The corner's position cannot be set without accessories.

(D) Only east–west distances need to be measured.

68. A retracement survey locates the east $\frac{1}{4}$ corners of Sections 7 and 18, the northwest corner of Section 18, and the north $\frac{1}{4}$ corner of Section 17 in a standard government township. All section lines in the township are shown on the plat as being 80 chains in length and at cardinal directions. What method should be used to set the lost northeast corner of Section 18?

(A) chain 80 chains east from the north $\frac{1}{4}$ corner of Section 18

(B) single proportional measurement

(C) double proportional measurement

(D) measure 40 chains south from the east $\frac{1}{4}$ corner of Section 7

69. In the illustrated recorded plat for Turkey Wing Subdivision, a survey locates no interior lot corners fronting on Wing Road, but finds the original front block corners 500.80 ft apart. No evidence of unwritten rights is found. At what distance from the block corners should the adjacent lot corners be set?

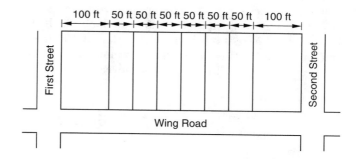

(A) 100.00 ft

(B) 100.16 ft

(C) 100.20 ft

(D) 100.40 ft

70. Why is the answer for Question 69 correct?

(A) The excess should be placed in the corner remanent lots.

(B) The excess is distributed equally in each lot.

(C) The excess is distributed by proportional measurement.

(D) The plat dimensions control remonumentation.

71. The subdivision plat shows Block C with a frontage of 500 ft. Four lots are shown with a frontage of 100 ft each; the end lot is not dimensioned, but scales 95 ft.

A survey finds a length between original block corners of 505 ft and no interior lot corners. At what distance from the block corner should the monument for the non-dimensioned lot be set?

(A) 95 ft

(B) 100 ft

(C) 101 ft

(D) 105 ft

72. Why is the answer for Question 71 correct?

(A) The intent was that all lots have a 100-ft frontage.

(B) Where an end lot is not dimensioned, excess is given to that lot.

(C) Proportional measurement provides the location.

(D) Where a lot is not dimensioned, scaled frontages govern.

73. A lot is 200 ft wide and 400 ft deep, with a constant slope from back to front. The ground elevation is 100 ft at the front of the lot and 104.5 ft at the back. If the entire lot is to be leveled to an elevation of 100 ft for building construction, approximately how many cubic yards of material will have to be removed?

(A) 6667 yd³

(B) 13,333 yd³

(C) 160,000 yd³

(D) 180,000 yd³

74. Given the following equation, what is the value of x?

$$20x^2 + 14x - 6 = 0$$

(A) −1 or 0.3

(B) 1 or 1.3

(C) 2 or 3

(D) 2.5

75. Refraction has which of the following effects on light rays passing through the atmosphere?

(A) bends rays toward the earth

(B) bends rays away from the earth

(C) reflects the rays

(D) has no effect

76. Given the illustrated triangle, what is the length of line AB?

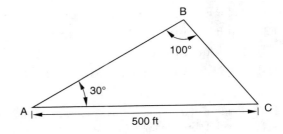

(A) 377.9 ft

(B) 382.9 ft

(C) 388.9 ft

(D) 389.9 ft

77. Which of the following is the best description of a professional surveyor's responsibility when testifying as an expert witness?

(A) to ensure that the needs of the client are met in the surveyor's testimony

(B) to be objective and truthful

(C) to provide direct testimony, but not answer questions under cross examination

(D) to carefully choose words and phrases to bolster the surveyor's image as an expert

78. Which is the correct expression of the number 0.00563 in scientific notation?

(A) 5.63×10^{-3}

(B) 5.63×10^{-2}

(C) 5.63×10^2

(D) 5.63×10^3

79. Given a lot forming a right triangle with a base of 80 ft and a hypotenuse of 100 ft, what is the length of the other side?

(A) 60 ft

(B) 62.5 ft

(C) 80 ft

(D) 128.06 ft

80. For state plane coordinate systems, Lambert projections are normally used in which of the following situations?

(A) for zones that have their long axis oriented east–west

(B) for zones that have their long axis oriented north–south

(C) for zones in higher latitudes

(D) for zones with little vertical relief

81. What is the latitude and departure for a traverse course of 245.10 ft at a bearing of N54°00′E?

(A) 144.07 ft, 198.29 ft

(B) 154.25 ft, 190.48 ft

(C) 190.48 ft, 154.25 ft

(D) 198.29 ft, 144.07 ft

82. A closed traverse has beginning coordinates in feet of 1000 and 1000, and ending coordinates of 988.25 and 1000.25. What is the direction and distance of the linear error of closure?

(A) N1°13′08″E, 11.75 ft

(B) N1°13′08″W, 11.75 ft

(C) N1°13′08″E, 12.25 ft

(D) N1°13′08″W, 12.25 ft

83. When applying the compass rule to the traverse described in Question 82, with a total traverse length of 2156.78 ft, what is the correction in latitude and departure, in feet, to be applied to a course with a measured length of 245.10 ft?

(A) 0.02, 1.13

(B) 0.03, 1.34

(C) 1.34, 0.03

(D) 1.45, 0.04

84. If the magnetic bearing of a line is S85°15′W at a certain location when the declination is 8°30′E, what is the true bearing?

(A) S76°45′W

(B) N86°15′W

(C) S86°15′W

(D) N88°15′W

85. A point observed through a level telescope appears to be higher (in relation to the level line of sight) than it actually is because of which of the following?

(A) parallax

(B) refraction

(C) curvature of the earth

(D) all of the above

STOP!

Do not continue on. This concludes the Morning Section of the examination. If you finish before time is called, you may check your work on this section of the examination. You may not turn to the Afternoon Section of the exam until you are told to do so by your proctor.

Name: _____

Last   First   Middle
                                       Initial

Instructions

This is a "closed book" examination. Subject to rules established by your state, you may not use any reference material. You may use any approved, battery-powered, silent calculator. No writing tablets or loose papers are permitted. Sufficient room for scratch work is provided in this examination booklet. You are not permitted to share or exchange materials with other examinees.

This section of the exam consists of 85 problems, each worth one point. You will have four hours in which to complete this section. Your score will be determined by the number of problems you solve correctly. No points will be deducted for incorrect answers. It is to your advantage to answer every question.

When permission has been given by your proctor, break the seal on the examination booklet and remove the answer sheet. Write your name immediately in the space indicated. Check that all pages are present and legible. If any part of this booklet is missing, your proctor will issue a new booklet.

All solutions must be entered on the answer sheet. No credit will be given for answers appearing in the examination booklet. Mark your answers with a No. 2 pencil. Do not use a pen. Marks must be dark and completely fill the "bubble." Record only one answer per problem; if you mark more than one answer, you will not receive credit for the problem. If you change an answer, be sure the old bubble is erased completely; incomplete erasures may be read as intended answers.

If you finish early, check your work and make sure you have correctly followed all instructions. After checking your answers, you may turn in your examination booklet and answer sheet and leave the examination room. Once you leave, you will not be permitted to return to work on your examination.

WAIT FOR PERMISSION TO BEGIN.

LAND SURVEYOR-IN-TRAINING SAMPLE EXAMINATION

AFTERNOON SECTION

PART 2

LAND SURVEYOR-IN-TRAINING
SAMPLE EXAMINATION

AFTERNOON SECTION

Name: _____

86 Ⓐ Ⓑ Ⓒ Ⓓ   103 Ⓐ Ⓑ Ⓒ Ⓓ   120 Ⓐ Ⓑ Ⓒ Ⓓ   137 Ⓐ Ⓑ Ⓒ Ⓓ   154 Ⓐ Ⓑ Ⓒ Ⓓ
87 Ⓐ Ⓑ Ⓒ Ⓓ   104 Ⓐ Ⓑ Ⓒ Ⓓ   121 Ⓐ Ⓑ Ⓒ Ⓓ   138 Ⓐ Ⓑ Ⓒ Ⓓ   155 Ⓐ Ⓑ Ⓒ Ⓓ
88 Ⓐ Ⓑ Ⓒ Ⓓ   105 Ⓐ Ⓑ Ⓒ Ⓓ   122 Ⓐ Ⓑ Ⓒ Ⓓ   139 Ⓐ Ⓑ Ⓒ Ⓓ   156 Ⓐ Ⓑ Ⓒ Ⓓ
89 Ⓐ Ⓑ Ⓒ Ⓓ   106 Ⓐ Ⓑ Ⓒ Ⓓ   123 Ⓐ Ⓑ Ⓒ Ⓓ   140 Ⓐ Ⓑ Ⓒ Ⓓ   157 Ⓐ Ⓑ Ⓒ Ⓓ
90 Ⓐ Ⓑ Ⓒ Ⓓ   107 Ⓐ Ⓑ Ⓒ Ⓓ   124 Ⓐ Ⓑ Ⓒ Ⓓ   141 Ⓐ Ⓑ Ⓒ Ⓓ   158 Ⓐ Ⓑ Ⓒ Ⓓ
91 Ⓐ Ⓑ Ⓒ Ⓓ   108 Ⓐ Ⓑ Ⓒ Ⓓ   125 Ⓐ Ⓑ Ⓒ Ⓓ   142 Ⓐ Ⓑ Ⓒ Ⓓ   159 Ⓐ Ⓑ Ⓒ Ⓓ
92 Ⓐ Ⓑ Ⓒ Ⓓ   109 Ⓐ Ⓑ Ⓒ Ⓓ   126 Ⓐ Ⓑ Ⓒ Ⓓ   143 Ⓐ Ⓑ Ⓒ Ⓓ   160 Ⓐ Ⓑ Ⓒ Ⓓ
93 Ⓐ Ⓑ Ⓒ Ⓓ   110 Ⓐ Ⓑ Ⓒ Ⓓ   127 Ⓐ Ⓑ Ⓒ Ⓓ   144 Ⓐ Ⓑ Ⓒ Ⓓ   161 Ⓐ Ⓑ Ⓒ Ⓓ
94 Ⓐ Ⓑ Ⓒ Ⓓ   111 Ⓐ Ⓑ Ⓒ Ⓓ   128 Ⓐ Ⓑ Ⓒ Ⓓ   145 Ⓐ Ⓑ Ⓒ Ⓓ   162 Ⓐ Ⓑ Ⓒ Ⓓ
95 Ⓐ Ⓑ Ⓒ Ⓓ   112 Ⓐ Ⓑ Ⓒ Ⓓ   129 Ⓐ Ⓑ Ⓒ Ⓓ   146 Ⓐ Ⓑ Ⓒ Ⓓ   163 Ⓐ Ⓑ Ⓒ Ⓓ
96 Ⓐ Ⓑ Ⓒ Ⓓ   113 Ⓐ Ⓑ Ⓒ Ⓓ   130 Ⓐ Ⓑ Ⓒ Ⓓ   147 Ⓐ Ⓑ Ⓒ Ⓓ   164 Ⓐ Ⓑ Ⓒ Ⓓ
97 Ⓐ Ⓑ Ⓒ Ⓓ   114 Ⓐ Ⓑ Ⓒ Ⓓ   131 Ⓐ Ⓑ Ⓒ Ⓓ   148 Ⓐ Ⓑ Ⓒ Ⓓ   165 Ⓐ Ⓑ Ⓒ Ⓓ
98 Ⓐ Ⓑ Ⓒ Ⓓ   115 Ⓐ Ⓑ Ⓒ Ⓓ   132 Ⓐ Ⓑ Ⓒ Ⓓ   149 Ⓐ Ⓑ Ⓒ Ⓓ   166 Ⓐ Ⓑ Ⓒ Ⓓ
99 Ⓐ Ⓑ Ⓒ Ⓓ   116 Ⓐ Ⓑ Ⓒ Ⓓ   133 Ⓐ Ⓑ Ⓒ Ⓓ   150 Ⓐ Ⓑ Ⓒ Ⓓ   167 Ⓐ Ⓑ Ⓒ Ⓓ
100 Ⓐ Ⓑ Ⓒ Ⓓ   117 Ⓐ Ⓑ Ⓒ Ⓓ   134 Ⓐ Ⓑ Ⓒ Ⓓ   151 Ⓐ Ⓑ Ⓒ Ⓓ   168 Ⓐ Ⓑ Ⓒ Ⓓ
101 Ⓐ Ⓑ Ⓒ Ⓓ   118 Ⓐ Ⓑ Ⓒ Ⓓ   135 Ⓐ Ⓑ Ⓒ Ⓓ   152 Ⓐ Ⓑ Ⓒ Ⓓ   169 Ⓐ Ⓑ Ⓒ Ⓓ
102 Ⓐ Ⓑ Ⓒ Ⓓ   119 Ⓐ Ⓑ Ⓒ Ⓓ   136 Ⓐ Ⓑ Ⓒ Ⓓ   153 Ⓐ Ⓑ Ⓒ Ⓓ   170 Ⓐ Ⓑ Ⓒ Ⓓ

TRIGONOMETRIC FORMULAS

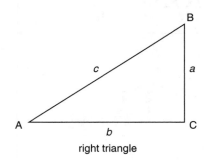

right triangle

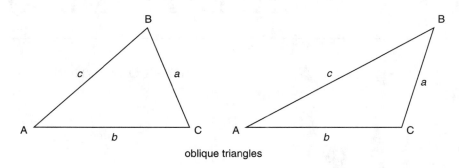

oblique triangles

Solution of Right Triangles

For angle A: $\sin A = \dfrac{a}{c}$ $\cos A = \dfrac{b}{c}$ $\tan A = \dfrac{a}{b}$ $\cot A = \dfrac{b}{a}$ $\sec A = \dfrac{c}{b}$ $\operatorname{cosec} A = \dfrac{c}{a}$

Given	Required		
a, b	A, B, c	$\tan A = \dfrac{a}{b} = \cot B$	$c = \sqrt{a^2 + b^2} = a\sqrt{1 + \dfrac{b^2}{a^2}}$
a, c	A, B, b	$\sin A = \dfrac{a}{c} = \cos B$	$b = \sqrt{(c+a)(c-a)} = c\sqrt{1 - \dfrac{a^2}{c^2}}$
A, a	B, b, c	$B = 90° - A$	$b = a \cot A$ $c = \dfrac{a}{\sin A}$
A, b	B, a, c	$B = 90° - A$	$a = b \tan A$ $c = \dfrac{b}{\cos A}$
A, c	B, a, b	$B = 90° - A$	$a = c \sin A$ $b = c \cos A$

Solution of Oblique Triangles

Given	Required			
A, B, a	b, c, C	$b = \dfrac{a \sin B}{\sin A}$	$C = 180° - (A + B)$	$c = \dfrac{a \sin C}{\sin A}$
A, a, b	B, c, C	$\sin B = \dfrac{b \sin A}{a}$	$C = 180° - (A + B)$	$c = \dfrac{a \sin C}{\sin A}$
a, b, C	A, B, c	$A + B = 180° - C$	$\tan\left(\dfrac{A - B}{2}\right) = \left(\dfrac{a - b}{a + b}\right)\left[\tan\left(\dfrac{A + B}{2}\right)\right]$	

$$c = \dfrac{a \sin C}{\sin A}$$

Given	Required			
a, b, c	A, B, C	$s = \dfrac{a + b + c}{2}$	$\sin\left(\dfrac{A}{2}\right) = \sqrt{\dfrac{(s - b)(s - c)}{bc}}$	

$$\sin\left(\dfrac{B}{2}\right) = \sqrt{\dfrac{(s - a)(s - c)}{ac}} \qquad C = 180° - (A + B)$$

Given	Required		
a, b, c	area	$s = \dfrac{a + b + c}{2}$	$\text{area} = \sqrt{s(s - a)(s - b)(s - c)}$
A, b, c	area	$\text{area} = \dfrac{bc \sin A}{2}$	
A, B, C, a	area	$\text{area} = \dfrac{a^2 \sin B \sin C}{2 \sin A}$	

HORIZONTAL CURVE FORMULAS

D = degree of curve, arc definition

$1°$ = 1-degree of curve

$2°$ = 2-degree of curve

PC = point of curve

PT = point of tangent

PI = point of intersection

I = intersection of angle;
angle between two tangents

L = length of curve
from PC to PT

T = tangent distance

E = external distance

R = radius

LC = length of long chord

M = length of middle ordinate

c = length of subchord

d = angle of subchord

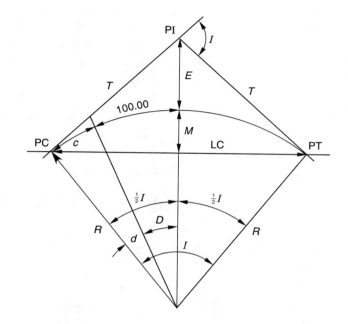

$$R = \frac{LC}{2\sin\left(\dfrac{I}{2}\right)} \qquad T = R\tan\left(\frac{I}{2}\right) = \frac{LC}{2\cos\left(\dfrac{I}{2}\right)}$$

$$\frac{LC}{2} = R\sin\left(\frac{I}{2}\right) \qquad D\,1° = R = 5729.58 \qquad D\,2° = \frac{5729.58}{2} \qquad D = \frac{5729.58}{R}$$

$$M = R\left[1 - \cos\left(\frac{I}{2}\right)\right] = R - R\cos\left(\frac{I}{2}\right)$$

$$\frac{E + R}{R} = \sec\left(\frac{I}{2}\right) \qquad \frac{R - M}{R} = \cos\left(\frac{I}{2}\right)$$

$$c = 2R\sin\left(\frac{d}{2}\right) \qquad d = \frac{c}{2R}$$

$$LC = 2R\sin\left(\frac{I}{2}\right) \qquad E = R\left[\sec\left(\frac{I}{2}\right) - 1\right] = R\sec\left(\frac{I}{2}\right) - R$$

MISCELLANEOUS CONVERSION FACTORS

1 meter = 3.280833 survey feet

1 survey foot = 0.3048006 meters

1 link = 0.66 feet

80 chains = 320 rods = 5280 feet

1 acre = 43,560 square feet

640 acres = 1 square mile

1 kilometer = 0.621370 mile

1 mile = 1.60935 kilometers

1 hectare = 2.47104 acres

$\pi = 3.14159$

1 radian = 57.2958 degrees

VERTICAL CURVE FORMULAS

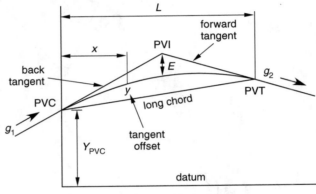

not to scale

L = length of curve

PVC = point of vertical curvature

PVI = point of vertical intersection

PVT = point of vertical tangency

g_1 = grade of back tangent

g_2 = grade of forward tangent

a = parabola constant

y = tangent offset

E = tangent offset at PVI

r = rate of change of grade

Y_{PVC} = elevation of PVC

$$y = ax^2 \qquad a = \frac{g_2 - g_1}{2L}$$

$$E = a\left(\frac{L}{2}\right)^2 \qquad r = \frac{g_2 - g_1}{L}$$

tangent elevation $= Y_{PVC} + g_1 x$

grade elvation $= Y_{PVC} + g_1 x + ax^2$

REFERENCE FORMULAS

Earthwork Formulas

$$\text{volume} = \frac{L(A_1 + A_2)}{2} \quad \text{[average end area formula]}$$

$$\text{volume} = \frac{L(A_1 + 4A_m + A_z)}{6} \quad \text{[prismoidal formula]}$$

$$\text{volume} = \frac{h(\text{area of base})}{3} \quad \text{[pyramid or cone]}$$

Area Formulas

$$\text{area} = \frac{X_A(Y_B - Y_N) + X_B(Y_C - Y_A) + X_C(Y_D - Y_B) + \cdots + X_N(Y_1 - Y_{N-1})}{2} \quad \text{[area by coordinates]}$$

$$\text{area} = w\left[\frac{h_1 + h_n}{2} + h_2 + h_3 + h_4 + \cdots + h_{n-1}\right] \quad \text{[trapezoidal rule]}$$

$$\text{area} = \frac{w[h_1 + 2(\sum h_{\text{odds}}) + 4(\sum h_{\text{evens}}) + h_n]}{3} \quad \text{[Simpson's one-third rule]}$$

Quadratic Equation Formula

$$ax^2 + bx + c = 0$$

$$x = \frac{-b \pm \sqrt{b^2 - 4ac}}{2a}$$

Correction Formulas

$$\text{correction for temperature} = C_t = 6.5 \times 10^{-6}(T - T_s)L$$

$$\text{correction for tension} = C_p = \frac{(P - P_s)L}{AE}$$

$$\text{correction for sag} = C_s = \frac{w^2 l^3}{24P^2}$$

T = temperature of tape during measurement, °F

T_s = temperature of tape during calibration, °F

L = distance measured, in ft

P = pull applied during measurement, in lb

P_s = pull applied during calibration, in lb

A = cross-section area of tape, in in^2

E = modulus of elasticity of tape, in psi

w = weight of tape, in lb per ft

l = length of unsupported span, in ft

Astronomy Formulas

$$\cos(\text{azimuth of sun}) = \frac{\sin\delta - \sin\phi\sin h}{\cos\phi\cos h}$$

$$Z = \text{bearing of Polaris at elongation} = \frac{p}{\cos\phi}$$

δ = declination angle

ϕ = latitude angle

h = altitude angle

p = polar distance of Polaris, angle

Photogrammetry Formulas

$$\text{average scale} = \frac{H - h_{\text{avg}}}{f}, \text{ in ft per in}$$

f = focal length of camera lens, in in

$H - h_{\text{avg}}$ = flying height above average ground surface, in ft

Stadia Formulas

$$\text{horizontal distance} = KS\cos^2\alpha$$

$$\text{vertical distance} = KS\sin\alpha\cos\alpha$$

K = stadia constant

S = rod intercept

α = vertical angle of sight

86. What is the area of the sector of a circle with a radius of 60 ft and a central angle of 60°?

    (A) 62.83 ft²

    (B) 1884.96 ft²

    (C) 5654.87 ft²

    (D) 11,309.73 ft²

87. What is the grade of a section of highway that has a centerline elevation of 114.50 ft at station 10+15.00 and a centerline elevation of 142.20 ft at station 22+25.00?

    (A) 1.3%

    (B) 2.3%

    (C) 2.8%

    (D) 3.3%

88. What is the area of the following quadrilateral?

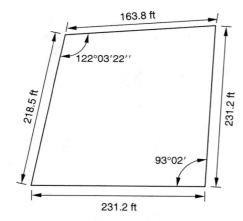

    (A) 39,842 ft²

    (B) 40,652 ft²

    (C) 41,856 ft²

    (D) 41,652 ft²

89. A 1000-foot-long earthen berm has a cross section with an area of 1500 ft² at one end and 2000 ft² at the other end, with area increasing at a constant rate. What is the approximate volume of material in the berm?

    (A) 55,556 yd³

    (B) 62,742 yd³

    (C) 64,815 yd³

    (D) 74,074 yd³

90. In the situation shown, which method could be used to locate Point C in relation to Points A and B, using only an EDM?

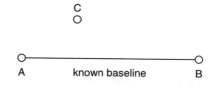

    (A) traverse

    (B) bearing-bearing intersection

    (C) triangulation

    (D) distance-distance intersection

91. Which of the following is the best definition of *mean sea level*?

    (A) a plane halfway between mean high water and mean low water

    (B) a plane halfway between mean higher high water and mean lower low water

    (C) National Geodetic Vertical Datum of 1929

    (D) the average of all the water heights over a tidal epoch

92. On a topographic map, which of the following is correct?

    (A) Contour lines crossing streams form V's that point upstream.

    (B) Contour lines crossing streams form V's that point downstream.

    (C) Contour lines crossing ridges form U's that point up the ridge.

    (D) Both (A) and (C) are true.

93. What is the effect on the location of a coastal boundary when natural and gradual changes occur to the shoreline?

    (A) The boundary remains at the location of the shoreline at the time of statehood.

    (B) The boundary remains at the location of the shoreline at the time of the original subdivision.

    (C) The boundary changes with the changing shoreline.

    (D) The boundary changes unless the state has a construction setback line in the area of the lot.

94. In the event of a conflict among the various elements in a deed description, which of the following calls would have the lowest priority?

(A) call for an artificial monument

(B) call for distance

(C) call for a natural monument

(D) call for an adjoiner with senior rights

95. In the event of a conflict among the various elements in a deed description, which of the following calls would have the second highest priority?

(A) call for an artificial monument

(B) call for distance

(C) call for a natural monument

(D) call for an adjoiner with senior rights

96. On a map developed from a Mercator projection (not a transverse Mercator), which of the following is true?

(A) Meridians converge.

(B) Parallels are arcs of circles.

(C) Meridians and parallels are at right angles.

(D) Parallels are uniformly spaced.

97. A horizontal highway curve is usually a portion of which curve?

(A) parabola

(B) circle

(C) hyperbola

(D) spiral

98. Choose the correct option for the underlined words in the following sentence: "He presented an explanation as to how temperature affects distance measurements and what effect this has on the accuracy of the survey."

(A) effects, affect

(B) affects, affect

(C) effects, effects

(D) no change is necessary

99. A state plane system grid distance is 4999.61 ft; the same distance measured on the face of the earth is 5000.00 ft. What is the scale factor? (Neglect sea level reduction.)

(A) 0.999922

(B) 0.999943

(C) 1.000025

(D) 1.000078

100. For a closed level loop beginning at an established benchmark and running through three new benchmarks, the resulting field elevations are as shown. What is the adjusted elevation for benchmarks NEW1, NEW2, and NEW3, respectively? All elevations are in feet.

Benchmark	Forward Run	Back Run
A195	100.00	99.92
	(published elevation)	
NEW1	102.50	102.38
NEW2	110.00	109.94
NEW3	113.00	113.00

(A) 102.44, 109.97, 113.00

(B) 102.45, 109.94, 112.92

(C) 102.48, 110.01, 113.04

(D) 102.50, 110.00, 113.00

101. For a closed traverse loop with measured angles and distances as shown, what is the ratio of closure after the angles are balanced? Assume a bearing of north for the course from Point 2 to Point 3. All distances are in meters.

BS	Setup	FS	Angle Right	Distance to FS
1	2	3	75°30′10″	380.00 m
2	3	4	120°40′30″	460.00 m
3	4	1	100°29′20″	555.06 m
4	1	2	63°20′20″	785.85 m

(A) 1:10,200

(B) 1:11,900

(C) 1:12,500

(D) 1:14,500

102. For the traverse in Question 101, what would be the adjusted latitude and departure from Point 2 to Point 3, using a compass rule adjustment? All distances are in meters.

(A) 379.993, −0.036

(B) 379.993, 0.025

(C) 380.010, −0.025

(D) 380.010, 0.036

103. What is the arc length of a horizontal curve with coordinates as shown and a radius of 1000.0 ft? All distances are in feet.

Point	Northing	Easting
PC	1000.00	1000.00
PI	1315.21	1181.99
PT	1439.69	1524.01

(A) 684.04 ft

(B) 691.24 ft

(C) 698.13 ft

(D) 1000.00 ft

104. What is the length of a 100-ft steel tape with a cross-sectional area of 0.003 in² and a modulus of elasticity of 30,000,000 lb/in² that is standard at a tension of 10 lb if it is pulled at a tension of 20 lb at standard temperature?

(A) 100.00 ft

(B) 100.01 ft

(C) 100.02 ft

(D) 100.11 ft

105. What is the length of side b in the illustrated triangle if the area is 43,560 ft²?

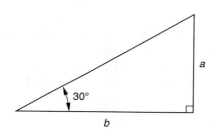

(A) 372.22 ft

(B) 279.53 ft

(C) 384.24 ft

(D) 388.45 ft

106. What is the station and offset of the southeast corner of the building shown?

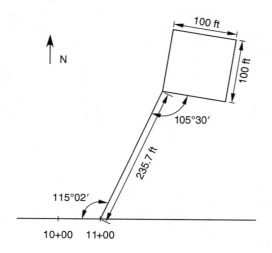

(A) 11+99.74, 213.56 ft

(B) 12+61.23, 191.21 ft

(C) 12+98.35, 197.00 ft

(D) 13+15.21, 185.29 ft

107. What is the length of the illustrated circular highway curve if the radius is 2000 ft?

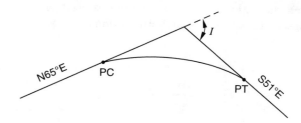

(A) 1596.25 ft

(B) 2000.00 ft

(C) 2234.02 ft

(D) 2524.96 ft

108. A retracement survey of a portion of a public land survey township results in the coordinates shown. The distances in the illustration are those provided in the original government survey, in chains (ch). What are the correct coordinates, in feet, for the lost southeast corner of Section 6? All coordinates are in feet.

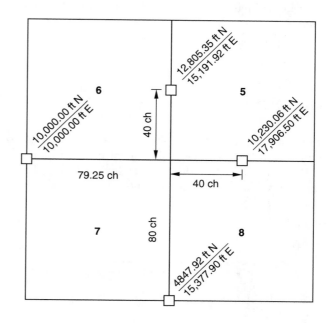

(A) 8826.64, 15,284.91

(B) 10,154.32, 15,256.65

(C) 10,152.87, 15,254.42

(D) 10,165.41, 15,258.24

109. At a primary tide station, the gauge reading of mean tide level is 5.25 m, and the long-term mean range of tide is 1.45 m. (Both values are from the latest National Tidal Datum Epoch.) Simultaneous tidal observations are conducted for one month between the primary station and a nearby point. The values listed are the mean values for the month of observation. What is the mean tidal range at the new station? Assume that the same ratio exists between the observed range and long-term mean range at both stations. All values are in meters.

	Monthly Mean Values	
	High Water	Low Water
Primary	5.65	4.05
New Station	3.29	1.89

(A) 1.27 m
(B) 1.40 m
(C) 1.45 m
(D) 1.60 m

110. A level telescope is set up at an elevation of 100.00 ft. At a distance of 4.5 mi, approximately what height above a point at elevation 95.00 ft would the telescope sight? The effect of curvature of the earth, in feet, may be estimated as $0.667 M^2$, while the effect of refraction may be estimated as $0.093 M^2$. M is the distance in miles.

(A) 5.0 ft
(B) 6.9 ft
(C) 16.6 ft
(D) 18.5 ft

111. In a peg check using the set-up shown, it is determined that the level has a significant collimation error. If the rod reading for the short sight is 4.12 ft, what is the rod reading for the long sight?

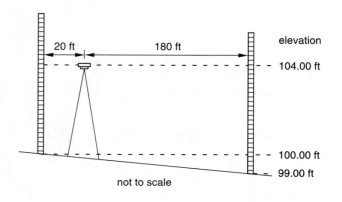

not to scale

(A) 4.92 ft
(B) 5.08 ft
(C) 5.12 ft
(D) 6.08 ft

112. A solar observation for azimuth determination using the altitude method produces the following data. What is the astronomic azimuth to the target?

latitude: 38°10′06″N
longitude: 84°15′24″W
observed angle right (target BS, sun FS): 235°44′52″
observed zenith angle: 55°56′06″
declination: +16°31′54″

(A) 028°49′29″
(B) 034°03′54″
(C) 140°49′29″
(D) 264°34′21″

113. A total station with a height of instrument of 5.00 ft, is set up over a point with an elevation of 100.00 ft. A slope distance measurement of 450.00 ft is made to a target with a zenith angle of 85°. What is the elevation of the target?

(A) 65.78 ft
(B) 139.22 ft
(C) 140.25 ft
(D) 144.22 ft

114. What is the mean and standard deviation of the following angle measurements?

Obs. No.	Angle	Obs. No.	Angle
1	31°02′29.3″	11	31°02′24.1″
2	31°02′24.0″	12	31°02′26.2″
3	31°02′27.9″	13	31°02′30.1″
4	31°02′26.8″	14	31°02′29.7″
5	31°02′26.1″	15	31°02′24.1″
6	31°02′25.9″	16	31°02′26.2″
7	31°02′26.1″	17	31°02′27.1″
8	31°02′27.8″	18	31°02′24.9″
9	31°02′27.2″	19	31°02′25.7″
10	31°02′28.0″	20	31°02′25.2″

(A) 31°02′25.8″, 2.5″
(B) 31°02′26.3″, 2.2″
(C) 31°02′26.6″, 1.7″
(D) 31°02′26.6″, 3.2″

115. A 200-foot-long parabolic vertical curve has equal tangents, a PI at station 18+00, a beginning grade of 1.25%, and an ending grade of −2.75%. The elevation of the PI is 270.19 ft. What is the elevation of the PVC?

(A) 257.69 ft

(B) 266.21 ft

(C) 268.94 ft

(D) 271.44 ft

116. What ground area would be covered by a five-time enlargement of a 24-inch by 36-inch negative at a scale of 1 in = 2000 ft?

(A) 3174 ac

(B) 4959 ac

(C) 5131 ac

(D) 79,339 ac

117. In the illustration, the interior dimensions are from the recorded subdivision plat while the coordinates are from a current resurvey where only the block corners were found. At what coordinates should a monument for corner A be set? All distances and coordinates are in feet.

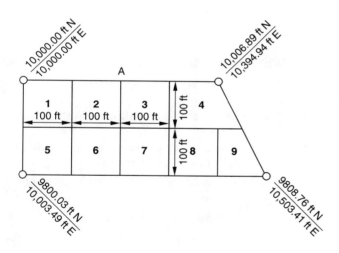

(A) 10,003.44, 10,197.47

(B) 10,003.44, 10,200.00

(C) 10,003.49, 10,198.90

(D) 10,003.49, 10,199.97

118. What would be the area of a tract of land described as the north one-half of Lot 12 as illustrated?

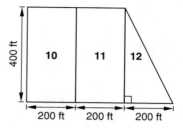

(A) 10,000 ft²

(B) 12,500 ft²

(C) 18,700 ft²

(D) 20,000 ft²

119. The following field notes represent a cross section of a levee. What is the area of the cross section as determined by the trapezoidal rule? All measurements are in feet.

Field Notes

<u>BS</u>   5.24   BM A (elevation = 100.00 ft)

FS	Offset
14.1	0
8.3	50
2.4	100
9.6	150
10.2	200
14.1	250

(A) 1052.5 ft²

(B) 1290.2 ft²

(C) 1295.0 ft²

(D) 1311.2 ft²

120. In the illustration, the interior dimensions are from the recorded plat, while the coordinates are from a recent resurvey. At what coordinates should the monument for corner A be set? All distances and coordinates are in feet.

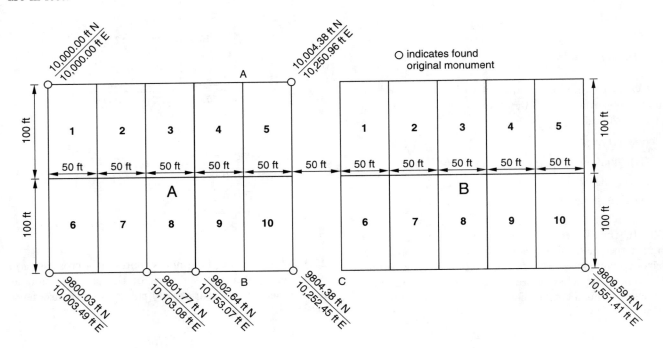

- (A) 10,003.49, 10,199.97
- (B) 10,003.50, 10,200.00
- (C) 10,003.50, 10,200.77
- (D) 10,004.21, 10,200.77

121. In Question 120, at what coordinates should the monument for corner B be set?

- (A) 9803.48, 10,202.22
- (B) 9803.49, 10,201.76
- (C) 9803.51, 10,202.66
- (D) 9803.51, 10,202.76

122. In Question 120, at what coordinates should the monument for corner C be set?

- (A) 9805.25, 10,302.24
- (B) 9805.25, 10,302.34
- (C) 9805.25, 10,302.40
- (D) 9805.25, 10,302.44

123. Which of the following is an example of parol evidence?

- (A) a scribed bearing tree called for in the original field notes
- (B) a statement from a former resident as to the location of an original corner
- (C) a measurement to a stream bank called for in the original field notes
- (D) a monument known to be replacing an original corner

124. Where would you look for information on a court case cited as "112 So. 274?"

- (A) in a federal reporter
- (B) in a regional reporter
- (C) in a state statute book
- (D) in the *American Law Reports*

125. Which of the following agencies would be the most appropriate source for maps identifying areas within 100-year flood zones?

(A) National Oceanic and Atmospheric Administration

(B) Corps of Engineers

(C) U.S. Geological Survey

(D) Federal Emergency Management Administration

126. Which of the following describes a design storm for calculating runoff quantities?

(A) the largest rainfall over a specific period of time

(B) the rainfall with the greatest duration over a specific period of time

(C) the average rainfall over a specific period of time

(D) a hypothetical rainfall representing the largest quantity of rain that statistically would be expected to fall over a specific period of time

127. Using a 24-in by 36-in sheet with a 1-inch margin, what is the best scale to depict a square tract of land with an area of 1.5 ac?

(A) 1:120

(B) 1:200

(C) 1:240

(D) 1:360

128. Choose the correct option for the underlined words in the following sentence: "Even though she had studied Russian in college, she could not understand native Russian <u>speakers they</u> all spoke too fast."

(A) speakers, they

(B) speakers. They

(C) speaking they

(D) no change is necessary

129. Given the illustrated plans for construction of a sewer line, what cut should be marked on the grade stake at station 11+00 if the ground elevation is 115.60 ft?

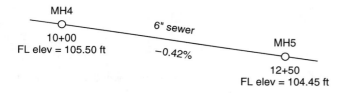

(A) 8.51

(B) 9.68

(C) 10.52

(D) 11.52

130. A level is placed on line between two benchmarks known to be at exactly the same elevation. The distance between the level and the near benchmark is one-third the distance to the farther benchmark. If the rod reading (in feet) is 4.422 at the near mark and 4.467 at the far mark, to what reading should the level be adjusted?

(A) 4.400

(B) 4.422

(C) 4.437

(D) 4.444

131. Which of the following would best describe *dedication*?

(A) recognition by the county of the completion of streets

(B) acquisition of property without the consent of the owner

(C) a means of transfer of rights to property

(D) a method of acquiring right to use of property by open occupation

132. Based on the field notes provided for cross sections, what is the ground elevation at a point 18 ft right of centerline at station 12+00? All units are in feet.

Field Notes

Station	+ HI	− Elevation	C/L						
BM A		44.90							
	3.30								
11+00				<u>4.4</u> <u>7.1</u> <u>4.9</u> 5.2 <u>5.1</u> <u>9.1</u> <u>11.2</u>					
				50 15 12 12 20 50					
12+00				<u>2.0</u> <u>1.3</u> <u>4.5</u> 5.0 <u>4.9</u> <u>5.7</u> <u>8.0</u>					
				50 18 12 12 18 50					
BM B	5.15	43.05							

(A) 42.5 ft

(B) 46.9 ft

(C) 49.5 ft

(D) 53.9 ft

133. In the following BASIC program, what is the value of C?

```
20 A = 45
30 REM A IS ANGLE IN DEGREES; B IS ANGLE
IN RADIANS
40 B = A * 0.01745
50 C = (105 * SIN(B))/2
```

(A) 0.72

(B) 12.15

(C) 37.12

(D) 74.25

134. Which of the following agencies would be the best source for maps depicting soil types in a given area?

(A) U.S. Department of Agriculture

(B) U.S. Geological Survey

(C) Federal Emergency Management Administration

(D) U.S. Coast & Geodetic Survey

135. What is the approximate north–south distance shown on a USGS 1:24,000 scale topographic map?

(A) 40,000 ft

(B) 45,000 ft

(C) 79,000 ft

(D) 91,000 ft

136. Choose the correct option for the underlined words in the following sentence: "Surveyors must now be registered to practice in all states, however, this has not always been the case."

(A) states; however, this

(B) states however, this

(C) states: however, this

(D) no change is necessary

137. Which of the following were generally not marked in the field during most original public land surveys?

(A) quarter-quarter section corners

(B) government lot lines

(C) quarter section corners

(D) both (A) and (B)

138. Which of the following types of GPS signal measurement results in the most precise measurements for static positioning?

(A) pseudorange

(B) code phase

(C) integrated Doppler

(D) carrier phase

139. What is the mean value for the angular observations tabulated below?

$$30°20'55''$$
$$30°21'02''$$
$$30°20'59''$$
$$30°20'50''$$
$$30°20'58''$$
$$30°20'59''$$
$$30°20'54''$$
$$30°20'56''$$
$$30°21'00''$$

(A) 30°20′44″

(B) 30°20′57″

(C) 30°20′58″

(D) 30°20′59″

140. What is the median value for the observations in Question 139?

(A) 30°20′44″

(B) 30°20′57″

(C) 30°20′58″

(D) 30°20′59″

141. What is the mode of the observations in Question 139?

(A) 30°20′44″

(B) 30°20′57″

(C) 30°20′58″

(D) 30°20′59″

142. What is the standard deviation for the observations in Question 139?

(A) 3.43″

(B) 3.54″

(C) 3.62″

(D) 4.80″

143. In the following field notes, what is the direct angle for the second set of angles?

(A) 157°41′03″

(B) 202°18′58″

(C) 270°05′23″

(D) 337°40′38″

Field Notes

At 2, BS 1:

Target	D/R	Circle Reading
1	D	000°00′34″
3	D	157°41′41″
3	R	337°40′38″
1	R	179°59′32″
1	R	270°05′23″
3	R	067°46′25″
3	D	247°47′24″
1	D	090°06′21″

144. Referring to the field notes in Question 143, what is the difference between the means of angles for each set of angles?

(A) 1″

(B) 2″

(C) 4″

(D) 6″

145. Referring to the field notes in Question 143, if the survey is to meet Third Order, Class II FGCC specifications, what would be the rejection limit for the mean of each set from the mean of all sets?

(A) 2″

(B) 4″

(C) 5″

(D) 20″

146. What is the decimal value of the binary number 110010?

(A) 11.0010

(B) 19

(C) 50

(D) 1100.10

147. Which of the following is a method of minimizing error due to imperfect adjustment of a level?

(A) double rodding

(B) three-wire leveling

(C) using longer sight lengths

(D) balancing sight lengths

148. The field notes of an old survey indicate that a line was run on a magnetic bearing of S85°15′W. Other research indicates that the declination for that area at the time of the survey was 8°30′E. What is the true bearing of the line?

(A) S76°45′W

(B) S81°00′W

(C) N86°15W

(D) S89°30′W

149. What is the maximum decimal value that can be stored in a computer byte?

(A) 63

(B) 127

(C) 255

(D) 511

150. To accomplish a large surveying project, it is necessary to borrow $50,000 for 6 months to cover additional payroll, surveying supplies, and other mobilization costs. Assuming simple interest of 16 percent per annum, how much would be due at the end of the 6-month period?

(A) $8,000

(B) $52,000

(C) $54,000

(D) $58,000

151. Under the early Spanish system of measurement, a league was which of the following?

(A) a unit of linear measurement

(B) a unit of measurement of area

(C) both (A) and (B)

(D) neither (A) nor (B)

152. A project will require $4520 of direct labor for a company that has a total fringe benefit and administrative overhead factor of 150 percent on a direct labor basis. Assuming that a 12 percent operating margin is desired, what fee should be quoted for the project?

(A) $5062

(B) $11,300

(C) $11,842

(D) $12,656

153. In the following description, which of the following would be considered an example of a bounds?

Description

A parcel of land situated in Leon County, Florida, being part of Section 16, Township 2 North, Range 1 East, and being more particularly described as follows: Begin at an old railroad iron marking the southwest corner of said Section 16; then go N80°42'E for 542.0 ft to the westerly right-of-way of Highway 20; then go N2°35'W for 120.0 ft along said right-of-way; then go N50°15'W for 502.5 ft to an old iron pipe; then go S15°09'W for 547.8 ft to the point of beginning.

    (A)   N80°42'E

    (B)   542.0 ft

    (C)   westerly right-of-way of Highway 20

    (D)   There are no bounds in the description.

154. What is the area contained within the parcel described in Question 153?

    (A)   1.49 ac

    (B)   2.29 ac

    (C)   3.61 ac

    (D)   1.2 hectares

155. Federal Geodetic Control Commission specifications for geodetic leveling call for maximum section misclosure (in millimeters) for Second Order, Class II work to be $8 \times \sqrt{D}$, where D is the distance in kilometers. For a level run between benchmarks 1 mile apart to meet this standard, what is the allowable misclosure?

    (A)   8 mm

    (B)   10 mm

    (C)   11 mm

    (D)   14 mm

156. When used to depict relief on maps, how should hachures be drawn?

    (A)   perpendicular to the slope direction

    (B)   diagonal to the slope direction

    (C)   parallel to the slope direction

    (D)   at various directions relative to the slope direction

157. For an urban survey being performed in accordance with American Land Title Association/American Congress on Surveying and Mapping minimum requirements, a closed traverse has a total length of 3000 m. What is the allowable closure?

    (A)   12 cm

    (B)   0.20 m

    (C)   0.60 m

    (D)   1.20 m

158. Field notes for the original survey of Section 16 indicate that all line lengths were 80 chains and all quarter corners were set at 40 chains. A resurvey has found acceptable monuments as shown. What are the correct coordinates for the center of Section 16? All coordinates are in feet.

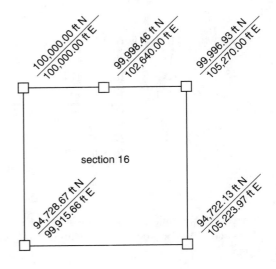

    (A)   97,360.24, 102,604.25

    (B)   97,361.94, 102,602.41

    (C)   97,361.94, 102,603.06

    (D)   97,361.94, 102,604.91

159. For the development of a geographic information system, you have scanned in all of the assessor plats for a small municipality. What would be the probable file structure for the resulting file?

    (A)   vector data structure

    (B)   metadata structure

    (C)   raster data structure

    (D)   none of the above

160. In the accompanying illustration, what is the stationing, in feet, of Point A?

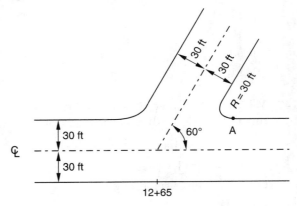

(A) 13+58.15
(B) 13+68.92
(C) 13+85.00
(D) 13+90.32

(A) 45 mph
(B) 50 mph
(C) 60 mph
(D) 70 mph

161. Refer to the illustration in Question 160. What is the distance from Point A to the point of intersection of the two centerlines?

(A) 103.92 ft
(B) 105.42 ft
(C) 106.18 ft
(D) 108.16 ft

162. A theodolite is set up 4.5 m above a point with an elevation of 52.60 m. The zenith angle at that point to a point at a distance of 1575 m is 75°25′. What is the elevation of the distant point?

(A) 195.24 m
(B) 350.26 m
(C) 462.37 m
(D) 466.87 m

163. What is the station, in feet, of the PT of the compound curve described here?

$$PC = 10+46.32$$
$$\Delta total = 68°00'$$
$$\Delta 1 = 35°00'$$
$$R1 = 600 \text{ ft}$$
$$R2 = 400 \text{ ft}$$

(A) 12+76.70
(B) 14+12.84
(C) 16+43.22
(D) 17+58.31

164. For the compound curve described in Question 163, what is the deflection angle (at the PC) for the PCC?

(A) 16°30′
(B) 17°30′
(C) 34°00′
(D) 35°00′

165. For what speed would a transitional spiral curve be designed with a length of 300 ft and a radius of 1200 ft? ($Ls = 1.6V^3/R$)

166. For the closed traverse illustrated, what is the length of side EA?

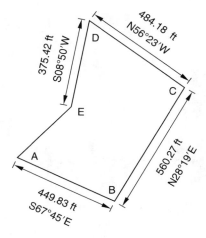

(A) 301.15 ft
(B) 309.26 ft
(C) 311.02 ft
(D) 312.00 ft

167. What is the area of the N$\frac{1}{2}$ of the NE$\frac{1}{4}$ of the NW$\frac{1}{4}$ of the SW$\frac{1}{4}$ of a typical public land survey section?

(A) 5 ac
(B) 10 ac
(C) 20 ac
(D) 40 ac

168. In which direction would the tract of land described in Question 167 lie in relation to the N$\frac{1}{2}$ of the NE$\frac{1}{4}$ of the NE$\frac{1}{4}$ of the SW$\frac{1}{4}$ of the same section?

(A) north
(B) east
(C) south
(D) west

169. What angular uncertainty is consistent with a linear measurement uncertainty of 1:10,000?

(A) 5″
(B) 10″
(C) 20″
(D) 30″

170. Using the linear and angular limits given in Question 169, what is the total spatial uncertainty for a traverse of 1500 ft in length?

   (A) ±0.10 ft

   (B) ±0.15 ft

   (C) ±0.20 ft

   (D) ±0.30 ft

ANSWER KEY FOR PARTS 1 AND 2

MORNING SECTION

1. B	18. D	35. B	52. C	69. B
2. A	19. D	36. A	53. A	70. C
3. C	20. B	37. A	54. C	71. D
4. C	21. B	38. B	55. A	72. B
5. D	22. C	39. C	56. C	73. A
6. C	23. D	40. D	57. D	74. A
7. B	24. A	41. C	58. B	75. A
8. C	25. A	42. D	59. A	76. C
9. D	26. A	43. D	60. A	77. B
10. D	27. C	44. A	61. A	78. A
11. A	28. D	45. C	62. C	79. A
12. A	29. C	46. D	63. B	80. A
13. C	30. C	47. D	64. B	81. A
14. D	31. A	48. B	65. D	82. B
15. B	32. C	49. C	66. C	83. C
16. C	33. A	50. A	67. A	84. B
17. B	34. B	51. B	68. C	85. B

AFTERNOON SECTION

86. B	103. C	120. C	137. D	154. C
87. B	104. B	121. D	138. D	155. B
88. C	105. D	122. D	139. B	156. C
89. C	106. C	123. B	140. C	157. B
90. D	107. C	124. B	141. D	158. D
91. D	108. C	125. D	142. A	159. C
92. A	109. A	126. D	143. A	160. B
93. C	110. C	127. B	144. C	161. D
94. B	111. D	128. B	145. C	162. D
95. C	112. A	129. C	146. C	163. C
96. C	113. D	130. A	147. D	164. B
97. B	114. C	131. C	148. C	165. C
98. D	115. C	132. A	149. C	166. D
99. A	116. A	133. C	150. C	167. A
100. C	117. D	134. A	151. C	168. D
101. D	118. D	135. B	152. D	169. C
102. B	119. C	136. A	153. C	170. D

SOLUTIONS

1.
$$\frac{x}{(14{,}000 \text{ ft})\left(12\,\dfrac{\text{in}}{\text{ft}}\right)} = \frac{1}{100{,}000}$$

$$x = \frac{(14{,}000 \text{ ft})\left(12\,\dfrac{\text{in}}{\text{ft}}\right)}{100{,}000}$$

$$= 1.68 \text{ in}$$

Answer is (B)

2.
$$\text{grid azimuth} = \text{astronomic azimuth}$$
$$- \text{convergence}$$
$$= 264°26'50'' - 00°26'42''$$
$$= 264°00'08''$$

Answer is (A)

3. Reliction is the exposure of land by the gradual lowering of water.

Answer is (C)

4.
$$\text{sum of the interior angles} = (n-2)(180°)$$
$$n = \text{number of angles}$$

With perfect closure,
$$\text{sum} = (5-2)(180°) = 540°$$

With misclosure of $10''$,
$$\text{sum} = 540° \pm 10''$$
$$= 540°00'10'' \text{ or } 539°59'50''$$

Answer is (C)

5. deflection angle at PI = interior angle

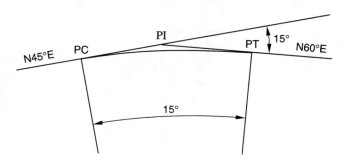

If tangent in $= $ N45°E, then
$$\text{tangent out} = \text{N45°E} + 15°$$
$$= \text{N60°E}$$

Answer is (D)

6. Using U.S. survey feet,
$$(350.25 \text{ m})\left(3.2808333\,\frac{\text{ft}}{\text{m}}\right) = 1149.11 \text{ ft}$$

Answer is (C)

7. Area (A) is equal to width (w) times length (l).
$$A = wl$$
$$w = \frac{A}{l}$$
$$= \frac{(1 \text{ ac})\left(43{,}560\,\dfrac{\text{ft}^2}{\text{ac}}\right)}{400 \text{ ft}}$$
$$= 108.9 \text{ ft}$$

Answer is (B)

8. The Domesday Book is the earliest of those mentioned. It was created by William the Conqueror in 1086 in an attempt to create a list of all land parcels in the whole of England, along with a listing of the land owner, acreage, tenants, land use, and livestock for each parcel. The other listed systems were all developed in the 1800s.

Answer is (C)

9.
$$(30 \text{ ch})\left(66\,\frac{\text{ft}}{\text{ch}}\right) = 1980 \text{ ft}$$
$$(30 \text{ ch})\left(4\,\frac{\text{rods}}{\text{ch}}\right) = 120 \text{ rods}$$

Answer is (D)

10. The hour angle method requires measurement of time and the horizontal angle of the sun. It does not require measurement of a vertical (or zenith) angle.

$$\boxed{\text{Answer is (D)}}$$

11. Using spherical coordinates, the declination of a celestial body is analogous to latitude using geographical coordinates.

$$\boxed{\text{Answer is (A)}}$$

12.
$$\frac{\text{angle nearest}}{\text{100-ft side}} = \arctan\left(\frac{\text{opposite side}}{\text{adjacent side}}\right)$$
$$= \arctan\left(\frac{200 \text{ ft}}{100 \text{ ft}}\right)$$
$$= 63°26'06''$$
$$\frac{\text{angle nearest}}{\text{200-ft side}} = \arctan\left(\frac{100 \text{ ft}}{200 \text{ ft}}\right)$$
$$= 26°33'54''$$

$$\boxed{\text{Answer is (A)}}$$

13. Mean is the arithmetic average; in this case, 29.24. Median is the middle number when numbers are arranged from lowest value to highest; in this case, 29.25. Mode is the most frequently occurring value; in this case, 29.25.

$$\boxed{\text{Answer is (C)}}$$

14. A patent is a conveyance from a sovereign power.

$$\boxed{\text{Answer is (D)}}$$

15.
$$\text{bearing} = \arctan\left(\frac{x_2 - x_1}{y_2 - y_1}\right)$$
$$= \arctan\left(\frac{300.75 - 100.00}{-150.25 - 100.00}\right)$$
$$= \arctan\left(\frac{200.75}{-250.25}\right)$$
$$= 38°44'$$

Since the value of the x-coordinate increases and the value of the y-coordinate decreases, the bearing is S38°44′E.

$$\text{distance} = \sqrt{(y_2 - y_1)^2 + (x_2 - x_1)^2}$$
$$= \sqrt{(-150.25 - 100.00)^2 + (300.75 - 100.00)^2}$$
$$= \sqrt{(-250.25)^2 + (200.75)^2}$$
$$= \sqrt{62,625.0625 + 40,300.5625}$$
$$= \sqrt{102,925.625}$$
$$= 320.82$$

$$\boxed{\text{Answer is (B)}}$$

16. Erosion is the gradual wearing away of upland by moving water.

$$\boxed{\text{Answer is (C)}}$$

17. In most states, the line of mean high water is the boundary between submerged land under navigable tidal water and bordering uplands.

$$\boxed{\text{Answer is (B)}}$$

18. The period used for a primary determination of average tidal heights in the U.S. is 19 years. That represents the period necessary for the varying relationships between the earth, sun, and moon to complete a full cycle (18.6 years), rounded to include an integral number of years.

$$\boxed{\text{Answer is (D)}}$$

19. When a negative with a scale of 1 in = 400 ft is enlarged four times, the scale of the enlargement is 1 in = 100 ft (400/4 = 100). For a ground distance of 1000 ft between targets, the photo distance is

$$\frac{1000 \text{ ft}}{100 \frac{\text{ft}}{\text{in}}} = 10 \text{ in}$$

$$\boxed{\text{Answer is (D)}}$$

20.

$$D = \frac{\$25,000 - \$5000}{5} = \frac{\$20,000}{5} = \$4000$$

Answer is (B)

21. Imagine a right triangle formed by a true vertical line, the 10° off-plumb rod, and the horizontal offset of the top of the rod from the vertical line.

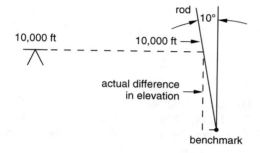

In a right triangle, by definition,

$$\text{cosine of an angle} = \frac{\text{adjacent side}}{\text{hypotenuse}}$$

Therefore, the actual difference in elevation is

$$(\cos 10°)(10.000 \text{ ft}) = 9.848 \text{ ft}$$
$$\text{elevation} = \text{HI} - \text{rod reading}$$
$$= 10.000 \text{ ft} - 9.848 \text{ ft}$$
$$= 0.152 \text{ ft}$$

Answer is (B)

22. By proportion,

$$\frac{\text{distance at 4000 ft}}{\text{distance at sea level}} = \frac{\text{radius at 4000 ft}}{\text{radius at sea level}}$$

$$\begin{aligned} \text{distance at} \\ \text{4000 ft} \end{aligned} = \frac{\begin{array}{c}(\text{distance at sea level}) \\ \times (\text{radius at 4000 ft})\end{array}}{\text{radius at sea level}}$$

$$= \frac{(2.5 \text{ mi})\left(5280 \dfrac{\text{ft}}{\text{mi}}\right)}{20,906,000 \text{ ft}}$$
$$= \frac{(13,200 \text{ ft})(20,910,000 \text{ ft})}{20,906,000 \text{ ft}}$$
$$= 13,202.53 \text{ ft}$$

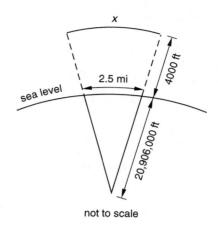

not to scale

Answer is (C)

23. For each square inch of photo,

$$\text{ground area} = \left(400 \frac{\text{ft}}{\text{in}}\right)\left(400 \frac{\text{ft}}{\text{in}}\right) = 160,000 \text{ ft}^2/\text{in}^2$$

For 20 in²,

$$A = (20 \text{ in}^2)\left(160,000 \frac{\text{ft}^2}{\text{in}^2}\right)$$
$$= 3,200,000 \text{ ft}^2$$
$$= 73.5 \text{ ac}$$

Answer is (D)

24. By definition, a radial line is perpendicular to the tangent at the PC.

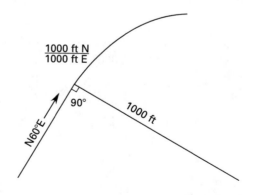

$$\text{bearing of radial from PC to RP} = \text{N60°E} + 90°$$
$$= \text{S30°E}$$
$$\text{latitude} = (\cos 30°)(1000 \text{ ft}) = -866.0 \text{ ft}$$
$$\text{departure} = (\sin 30°)(1000 \text{ ft}) = +500.0 \text{ ft}$$

coordinates of RP:

$$N = N \text{ of PC} + \text{latitude}$$
$$= 1000 \text{ ft} - 866.0 \text{ ft}$$
$$= 134.0 \text{ ft}$$

$$E = E \text{ of PC} + \text{departure}$$
$$= 1000 \text{ ft} + 500.0 \text{ ft}$$
$$= 1500.0 \text{ ft}$$

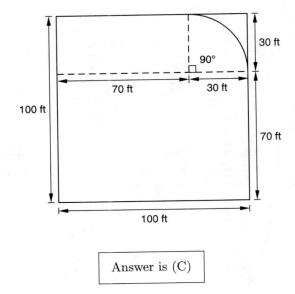

$$\boxed{\text{Answer is (A)}}$$

25. Due to the convergence of meridians, a unit of arc of longitude becomes increasingly smaller than a unit of arc of latitude as one goes north or south of the equator.

$$\boxed{\text{Answer is (A)}}$$

26. Due to the convergence of meridians, grid north is east of true north for points that are east of the central meridian and west of true north for points that are west of the central meridian.

$$\boxed{\text{Answer is (A)}}$$

27.

$$\begin{array}{l}\text{area within} \\ \text{circle segment}\end{array} = \left(\frac{90°}{360°}\right) \pi r^2$$
$$= \left(\frac{90°}{360°}\right) \pi (30 \text{ ft})^2$$
$$= 706.86 \text{ ft}^2$$

$$\begin{array}{l}\text{area within} \\ \text{remainder of lot}\end{array} = (70 \text{ ft})(100 \text{ ft}) + (30 \text{ ft})(70 \text{ ft})$$
$$= 7000 \text{ ft}^2 + 2100 \text{ ft}^2$$
$$= 9100.00 \text{ ft}^2$$
$$\text{total area} = 706.86 \text{ ft}^2 + 9100.00 \text{ ft}^2$$
$$= 9806.86 \text{ ft}^2$$

$$\boxed{\text{Answer is (C)}}$$

28. The controlling call is "to the shore of Cripple Creek." The creek is a natural monument, which has precedence over the calls for distance and bearing.

$$\boxed{\text{Answer is (D)}}$$

29. In government subdivisions, excess and deficiency is placed in lots in the northern and western tier of sections. Section 30 is in the western tier of sections. Since the south line of Section 30 is slightly less than a full 80 chains, the deficiency should be placed in the westernmost lot. Normally, that lot would have a width of 20 chains, with the remainder of the south line being 60 chains. With the deficiency, the lot width at the south line would be

$$79.60 \text{ ch} - 60.00 \text{ ch} = 19.60 \text{ ch}$$

$$\boxed{\text{Answer is (C)}}$$

30. Since the excess and deficiency for the section should be placed in the westernmost tier of lots, the length of the south line of the SE$\frac{1}{4}$ of the SW$\frac{1}{4}$ of Section 30 should be a full 20 chains.

$$\boxed{\text{Answer is (C)}}$$

31. Section 29 of the township is not on the northern or western tier. The lost quarter corner should be placed at the midpoint of the south line. The quarter-quarter corners should be placed at the midpoints of the two quarters. Therefore, the length of the south line of the $SW\frac{1}{4}$ of the $SW\frac{1}{4}$ should be

$$\frac{5278.20 \text{ ft}}{4} = 1319.55 \text{ ft}$$

Answer is (A)

32.

$$\text{area A} = wl = (150 \text{ ft})(600 \text{ ft}) = 90,000 \text{ ft}^2$$

$$\text{area B} = \tfrac{1}{2}wl = \frac{(50 \text{ ft})(300 \text{ ft})}{2} = 7500 \text{ ft}^2$$

$$\text{area C} = \tfrac{1}{2}wl = \frac{(150 \text{ ft})(300 \text{ ft})}{2} = 22,500 \text{ ft}^2$$

$$\text{total area} = 120,000 \text{ ft}^2$$

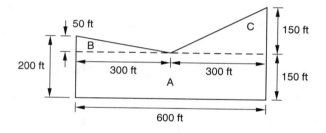

Answer is (C)

33. Dual-frequency GPS receivers collect data that can be used to model the ionospheric delay and thereby compensate for its effect on the signal. Therefore, they may be used for baselines of greater length and also require shorter observation times for comparable precision, as compared to single-frequency receivers.

Answer is (A)

34. Since the southeast corner of Section 31 is on the township exterior, single proportional measurement should be used to reestablish lost corners.

Answer is (B)

35. Polaris is located at the tip of the handle of Ursa Minor (the Little Dipper).

Answer is (B)

36. A Gunter's chain is 66 ft in length and is divided into 100 links. Therefore, each link is 0.66 ft or 7.92 in.

Answer is (A)

37. The thread readings are approximately

5.247 ft
5.130 ft
5.013 ft

$$\text{distance} = (\text{top} - \text{bottom thread})(\text{stadia constant})$$
$$= (5.247 \text{ ft} - 5.013 \text{ ft})(100)$$
$$= 23.4 \text{ ft}$$

Answer is (A)

38. Curvature of the earth would have the effect of making the rod reading greater, due to the constant dropping of the earth's surface from the horizontal level line.

Answer is (B)

39. Refraction is the bending of light rays and has the opposite effect from curvature. Therefore, it would make the rod reading less.

Answer is (C)

40. On a Lambert projection, meridians are shown as converging straight lines, while parallels are shown as uniformly spaced arcs of circles.

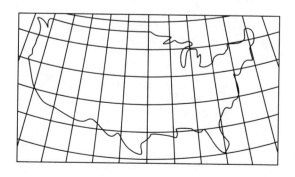

Answer is (D)

41. Since the southwest corner of Section 31 is the township corner, double proportional measurement should be used to reestablish the corner.

$$\boxed{\text{Answer is (C)}}$$

42. DMD stands for Double Meridian Distance and describes a method for determining the area of a closed figure.

$$\boxed{\text{Answer is (D)}}$$

43. The perpendicular bisector of the line of closure will point to the traverse station where the error was probably made.

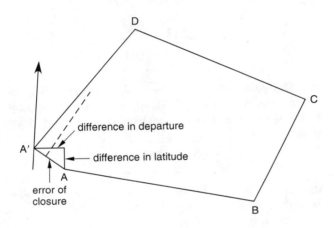

$$\boxed{\text{Answer is (D)}}$$

44. According to the Pythagorean theorem,

$$
\begin{aligned}
\text{error of closure} &= \sqrt{\Delta \text{ northing}^2 + \Delta \text{ easting}^2} \\
&= \sqrt{(0.37 \text{ ft})^2 + (0.24 \text{ ft})^2} \\
&= \sqrt{0.1369 \text{ ft}^2 + 0.0576 \text{ ft}^2} \\
&= \sqrt{0.1945 \text{ ft}^2} \\
&= 0.44 \text{ ft}
\end{aligned}
$$

ratio of error = 0.44:2181.14 or 1:4957

$$\boxed{\text{Answer is (A)}}$$

45. Geodetic elevations are usually referenced to the National Geodetic Vertical Datum (NGVD). Mean tide level and mean sea level are both local tidal data, while the North American Datum (NAD) 1927 is a horizontal datum.

$$\boxed{\text{Answer is (C)}}$$

46.
$$
\begin{aligned}
\text{arc length} &= \text{PT station} - \text{PC station} \\
&= (33+45.67)\text{ft} - (23+45.67)\text{ft} \\
&= 78.67 \text{ ft} - (-22.67 \text{ ft}) \\
&= 1000 \text{ ft}
\end{aligned}
$$

or

$$
\begin{aligned}
\text{arc length} &= \left(\frac{60°}{360°}\right) (\text{perimeter of circle}) \\
&= \left(\frac{60°}{360°}\right) (2\pi r) \\
&= \left(\frac{60°}{360°}\right) (2)\pi (954.93 \text{ ft}) \\
&= 1000 \text{ ft}
\end{aligned}
$$

$$\boxed{\text{Answer is (D)}}$$

47.
$$
\begin{aligned}
\text{surface distance} &= \frac{\text{grid distance}}{\text{scale factor}} \\
&= \frac{5280.00 \text{ ft}}{0.9999424} \\
&= 5280.30 \text{ ft}
\end{aligned}
$$

$$\boxed{\text{Answer is (D)}}$$

48.

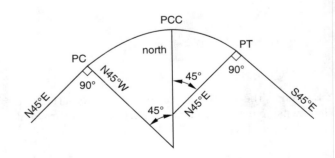

$$\boxed{\text{Answer is (B)}}$$

49. The highest elevation is somewhere between 150 ft and the next higher contour (160 ft). Therefore, 152 ft is the best answer.

Answer is (C)

50. The hachures indicate a depression. Therefore, the lowest elevation is somewhere between 50 ft and the next lower contour (40 ft). Therefore, 43 ft is the best answer.

Answer is (A)

51.
$$\Delta \text{ height} = \tan(90° - 87°)(5280 \text{ ft})$$
$$= 276.71 \text{ ft}$$
$$\text{elevation} = \text{HI} + \Delta \text{ height}$$
$$= 1000 \text{ ft} + 276.71 \text{ ft}$$
$$= 1276.71 \text{ ft}$$

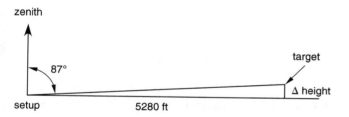

Answer is (B)

52. The process of obtaining easement rights from long-continued usage is called *prescription*.

Answer is (C)

53. The process of acquiring title to property by long-term occupation is called *adverse possession*.

Answer is (A)

54.
$$\text{baseline distance} = (50+41) - (32+42)$$
$$= 1799 \text{ ft}$$
$$\text{distance BC} = 42.12 \text{ ft} + 20.55 \text{ ft}$$
$$= 62.67 \text{ ft}$$

By Pythagorean theorem,
$$\text{distance AB} = \sqrt{\text{AC}^2 + \text{BC}^2}$$
$$= \sqrt{(1799 \text{ ft})^2 + (62.67 \text{ ft})^2}$$
$$= \sqrt{3,240,328.53 \text{ ft}^2}$$
$$= 1800.09 \text{ ft}$$

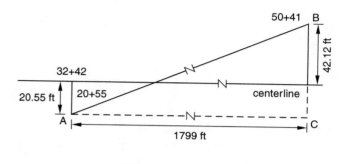

Answer is (C)

55.

	BS		FS	
Thread	Average		Thread	Average
5.242			6.813	
4.495	4.496		6.060	6.058
3.751			5.302	

$$\text{elevation FS} = \text{elevation BS} + \text{BS} - \text{FS}$$
$$= 100.000 \text{ ft} + 4.496 \text{ ft} - 6.058 \text{ ft}$$
$$= 98.438 \text{ ft}$$

Answer is (A)

56. Assuming a stadia constant of 100,
$$\text{distance} = (\text{top thread} - \text{bottom thread})(100)$$
$$= (5.242 \text{ ft} - 3.751 \text{ ft})(100)$$
$$= 149.1 \text{ ft}$$

Answer is (C)

57. The standard width of a time zone is 15°. This may be confirmed by realizing that the earth is divided into 24 time zones. Therefore, a time zone is
$$\frac{360°}{24} = 15°$$

Answer is (D)

58.

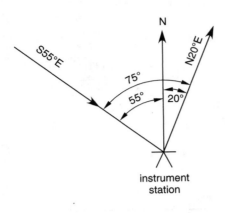

Answer is (B)

59. A vertical highway curve is usually a portion of a parabola.

Answer is (A)

60. On a plane projection, parallel to the equator and tangent to the North Pole, meridians converge at the pole.

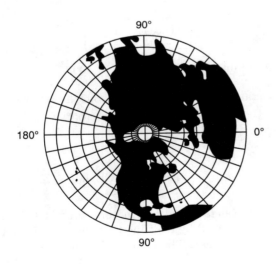

Answer is (A)

61.

$$\text{focal length} = 6 \text{ in} = 0.5 \text{ ft}$$

$$\text{scale} = \frac{\text{focal length}}{\text{altitude}}$$

$$\frac{1}{x} = \frac{0.5 \text{ ft}}{2400 \text{ ft}}$$

$$x = \frac{2400 \text{ ft}}{0.5 \text{ ft}}$$

$$x = 4800 \text{ ft}$$

$$\text{scale} = 1{:}4800$$

$$1 \text{ in} = \frac{4800 \text{ ft}}{12 \frac{\text{in}}{\text{ft}}} = 400 \text{ ft}$$

Answer is (A)

62. The senior deed called for one half of the lot. The north and south lines would be divided in half to set the monument.

$$\frac{198.5 \text{ ft}}{2} = 99.25 \text{ ft}$$

Answer is (C)

63. Senior rights control.

Answer is (B)

64. The nature of the two descriptions has created a hiatus, which may still belong to the original owner. The 1960 conveyance called for the western one-half. The 1990 conveyance called for the eastern 100 ft. The northwest corner of the 1990 conveyance would be 100 ft from the northeast corner of the original lot.

Answer is (B)

65. $$1 \text{ km} = 0.6 \text{ mi}$$

For state standards,

$$\text{maximum closure for } 1 \text{ km} = (0.05 \text{ ft})(\sqrt{0.6})$$
$$= 0.04 \text{ ft}$$

$$\frac{0.04\ \text{ft}}{3.2808\ \frac{\text{ft}}{\text{m}}} = 0.012\ \text{m}$$

$$= 12\ \text{mm}$$

State standards are equivalent to Third Order.

Answer is (D)

66. An obliterated corner is defined as a corner where there are no remaining traces of the original monument or its accessories, but whose location has been perpetuated by the acts or testimony of interested owners, competent surveyors, or local authorities.

Answer is (C)

67. The testimony of the landowner makes the location of the iron pipe acceptable.

Answer is (A)

68. Reestablishment of lost interior section corners should be accomplished by double proportional measurement.

Answer is (C)

69. Excess and deficiency should be proportioned within the block. The excess per frontage foot of lot is

$$\frac{0.80\ \text{ft}}{500\ \text{ft}} = 0.0016\ \text{ft}$$

For a 100-ft lot, the excess is

$$(100)(0.0016\ \text{ft}) = 0.16\ \text{ft}$$

The required lot corners should be set 100.16 ft from the block corner.

Answer is (B)

70. Excess between found original monuments in a subdivision is distributed by proportional measurement.

Answer is (C)

71. Where an end lot is not dimensioned, the excess and deficiency is given to that lot. The end lot dimension should be

505 ft − 400 ft (total of record lot sizes) = 105 ft

Answer is (D)

72. Where an end lot is not dimensioned, any excess and deficiency is given to that lot.

Answer is (B)

73. In the typical cross section shown,

$$\begin{aligned}
\text{area} &= \left(\tfrac{1}{2}\right)(\text{base})(\text{height}) \\
&= \left(\tfrac{1}{2}\right)(400\ \text{ft})(4.5\ \text{ft}) \\
&= 900\ \text{ft}^2
\end{aligned}$$

$$\begin{aligned}
\text{volume} &= (\text{cross-sectional area})(\text{width}) \\
&= (900\ \text{ft}^2)(200\ \text{ft}) = 180{,}000\ \text{ft}^3
\end{aligned}$$

$$\frac{180{,}000\ \text{ft}^3}{27\ \frac{\text{ft}^3}{\text{yd}^3}} = 6667\ \text{yd}^3$$

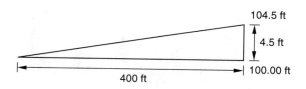

Answer is (A)

74. Factoring is the easiest method of solving this quadratic equation.

$$\begin{aligned}
20x^2 + 14x - 6 &= 0 \\
(2x + 2)(10x - 3) &= 0 \\
2x + 2 &= 0 \\
10x - 3 &= 0
\end{aligned}$$

$$x = \frac{-2}{2} = -1$$

$$x = \frac{3}{10} = 0.3$$

An alternate method is to use the quadratic equation.

$$x = \frac{-b \pm \sqrt{b^2 - 4ac}}{2a}$$

$$= \frac{-14 \pm \sqrt{(14)^2 - (4)(20)(-6)}}{(2)(20)}$$

$$= \frac{-14 \pm \sqrt{676}}{40}$$

$$= \frac{-14 + 26}{40} = 0.3$$

or

$$= \frac{-14 - 26}{40} = -1.0$$

Answer is (A)

75. Refraction in the earth's atmosphere causes light rays to bend downward toward the earth.

Answer is (A)

76.

$$\text{angle C} = 180° - (A + B)$$
$$= 180° - (100° + 30°)$$
$$= 50°$$

By law of sines,

$$\frac{AB}{\sin C} = \frac{AC}{\sin B}$$
$$AB = \frac{AC(\sin C)}{\sin B}$$
$$= \frac{(500 \text{ ft})(\sin 50°)}{\sin 100°}$$
$$= 388.9 \text{ ft}$$

Answer is (C)

77. An expert witness is required to be objective and truthful, regardless of the interests of the client or of personal interests.

Answer is (B)

78. In converting 0.00563 to scientific notation, count over three places to the right of the first significant figure. The correct notation is 5.63×10^{-3}.

Answer is (A)

79. By Pythagorean theorem,

$$(\text{hypotenuse})^2 = (\text{base})^2 + (\text{height})^2$$

$$\text{height} = \sqrt{(\text{hypotenuse})^2 - (\text{base})^2}$$
$$= \sqrt{(100 \text{ ft})^2 - (80 \text{ ft})^2}$$
$$= 60 \text{ ft}$$

Answer is (A)

80. Lambert projections are normally used in zones with their long axis oriented east–west.

Answer is (A)

81.

$$\text{latitude} = \cos(\text{bearing})(\text{length})$$
$$= (\cos 54°)(245.10 \text{ ft})$$
$$= 144.07 \text{ ft}$$
$$\text{departure} = \sin(\text{bearing})(\text{length})$$
$$= (\sin 54°)(245.10 \text{ ft})$$
$$= 198.29 \text{ ft}$$

Answer is (A)

82.

$$\text{error of closure} = \sqrt{\Delta \text{latitude}^2 + \Delta \text{departure}^2}$$
$$= \sqrt{(1000 \text{ ft} - 988.25 \text{ ft})^2 + (1000.25 \text{ ft} - 1000 \text{ ft})^2}$$
$$= \sqrt{(11.75 \text{ ft})^2 + (0.25 \text{ ft})^2}$$
$$= 11.75 \text{ ft}$$

$$\tan(\text{bearing}) = \frac{\Delta \text{ departure}}{\Delta \text{ latitude}}$$

$$= \frac{0.25}{11.75}$$

$$= 1°13'08''$$

By inspection of coordinates, the bearing is N1°13'08''W.

> Answer is (B)

83. Using the compass rule,

$$\begin{matrix}\text{correction for} \\ \text{latitude for} \\ \text{a course}\end{matrix} = \left(\frac{\text{length of course}}{\text{length of perimeter}}\right)\begin{pmatrix}\text{total error} \\ \text{in latitude}\end{pmatrix}$$

$$= \left(\frac{245.10 \text{ ft}}{2156.78 \text{ ft}}\right)(11.75 \text{ ft})$$

$$= 1.34 \text{ ft}$$

$$\begin{matrix}\text{correction for} \\ \text{departure for} \\ \text{a course}\end{matrix} = \left(\frac{\text{length of course}}{\text{length of perimeter}}\right)\begin{pmatrix}\text{total error} \\ \text{in departure}\end{pmatrix}$$

$$= \left(\frac{245.10 \text{ ft}}{2156.78 \text{ ft})}\right)(0.25 \text{ ft})$$

$$= 0.03 \text{ ft}$$

> Answer is (C)

84. When correcting magnetic bearings, add east declination and subtract west declination.

magnetic bearing	S85°15'W
magnetic azimuth	265°15'
declination	+8°30'
true azimuth	273°45'
true bearing	N86°15'W

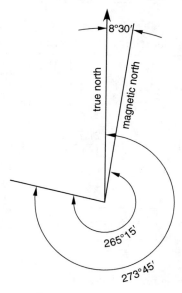

> Answer is (B)

85. Refraction is the bending of light rays toward the earth. This has the effect of making objects look higher than they actually are in relation to the level line of sight.

> Answer is (B)

86.

$$\text{area of circle} = \pi r^2$$

$$\text{area of the required sector} = \left(\frac{60°}{360°}\right)\pi(60 \text{ ft}^2)$$

$$= 1884.96 \text{ ft}^2$$

Answer is (B)

87.

Elevation	Station
114.50 ft	10+15.00
142.20 ft	22+25.00
Δ + 27.70 ft	12+10.00

$$\text{grade} = \frac{27.70 \text{ ft}}{1210.00 \text{ ft}}$$

$$= 0.023$$

$$= 2.3\%$$

Answer is (B)

88. area of triangle A $= \frac{1}{2}ab\sin C$

$$= (163.8 \text{ ft})(218.5 \text{ ft})$$

$$\times (\sin 122°03'22'')/2$$

$$= 15,166.65 \text{ ft}^2$$

area of triangle B $= \frac{1}{2}ab\sin C$

$$= (231.2 \text{ ft})(231.2 \text{ ft})(\sin 93°02')/2$$

$$= 26,689.27 \text{ ft}^2$$

total area $= 15,166.65 \text{ ft}^2 + 26,689.27 \text{ ft}^2$

$$= 41,855.93 \text{ ft}^2$$

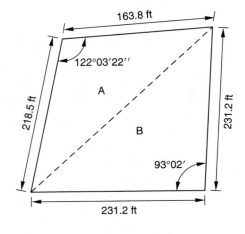

Answer is (C)

89.

$$V = l\left(\frac{A_1 + A_2}{2}\right)$$

$$= (1000 \text{ ft})\left(\frac{1500 \text{ ft} + 2000 \text{ ft}}{2}\right)$$

$$= 1,750,000 \text{ ft}^3$$

$$\frac{1,750,000 \text{ ft}^3}{\frac{27 \text{ ft}^3}{1 \text{ yd}^3}} = 64,815 \text{ yd}^3$$

Answer is (C)

90. A distance-distance intersection could be used to locate Point C in relation to the baseline AB, using only an EDM.

Answer is (D)

91. Mean sea level is a local datum and is the average of all the water heights over a 19-year tidal epoch.

Answer is (D)

92. On a topographic map, contour lines crossing streams form V's that point upstream.

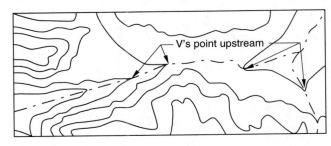

Answer is (A)

93. When natural and gradual changes occur to a shoreline (accretion or erosion), the coastal boundary changes with the changing shoreline.

Answer is (C)

94. In general, calls for distance should be given the lowest priority of the choices given.

Answer is (B)

95. Calls for natural monuments should be given the second highest priority of the choices given, after calls for adjoiners with senior rights.

> Answer is (C)

96. On a Mercator projection, meridians and parallels are shown as straight lines at right angles.

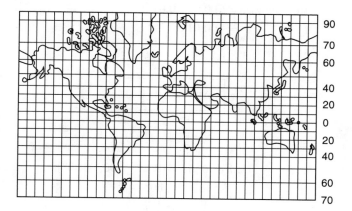

> Answer is (C)

97. A horizontal highway curve is usually a portion of a circle.

> Answer is (B)

98. No change is necessary. *Affect* is a verb meaning "to influence" or "to change." *Effect* is a noun meaning "result."

> Answer is (D)

99. grid distance = (surface distance) (scale factor)

Therefore,

$$\text{scale factor} = \frac{\text{grid distance}}{\text{surface distance}}$$

$$= \frac{4999.61 \text{ ft}}{5000.00 \text{ ft}}$$

$$= 0.999922$$

> Answer is (A)

100.

Forward Elevation	Δ	Back Elevation	Δ	Δ	Average Elevation
100.00		99.92			100.00
	+2.50		−2.46	+2.48	
102.50		102.38			102.48
	+7.50		−7.56	+7.53	
110.00		109.94			110.01
	+3.00		−3.06	+3.03	
113.00		113.00			113.04

> Answer is (C)

101.

Setup	Angle Measured	Angle Adjusted	Assumed Bearing from BS	Distance from BS	Latitude	Departure
2	075°30′10″	075°30′05″	N00°00′00″E	380.00	+380.000	0.000
3	120°40′30″	120°40′25″	N59°19′35″W	460.00	+234.668	−395.640
4	100°29′20″	100°29′15″	S41°09′40″W	555.06	−417.883	−365.329
1	063°20′20″	063°20′15″	S75°30′05″E	785.85	−196.743	+760.824
	360°00′20″	360°00′00″		2180.91	+0.042	−0.145

total length = 2180.91 m

closure: latitude = +0.042 m

departure = −0.145 m

$$\text{net} = \sqrt{0.042 \text{ m}^2 + 0.145 \text{ m}^2}$$

$$= 0.15 \text{ m}$$

ratio = 0.15 : 2180.91 = 1 : 14,539

> Answer is (D)

102.

$$\begin{aligned}\text{latitude} \atop \text{adjustment} &= \frac{\text{course distance}}{\text{total distance}} \text{ (latitude error)}\\ &= \left(\frac{380.00 \text{ m}}{2180.91 \text{ m}}\right)(0.042 \text{ m})\\ &= 0.007 \text{ m}\end{aligned}$$

$$\begin{aligned}\text{departure} \atop \text{adjustment} &= \frac{\text{course distance}}{\text{total distance}} \text{ (departure error)}\\ &= \left(\frac{380.00 \text{ m}}{2180.91 \text{ m}}\right)(0.145)\\ &= 0.025 \text{ m}\end{aligned}$$

Unadjusted		Adjustment		Adjusted	
Latitude	Departure	Latitude	Departure	Latitude	Departure
+380.000	0.000	−0.007	+0.025	379.993	0.025

Answer is (B)

103.

$$\begin{aligned}\text{tangent bearing} \atop \substack{\text{into curve} \\ \text{(PC to PI)}} &= \arctan\left(\frac{E}{N}\right)\\ &= \arctan\left(\frac{1181.99 \text{ ft} - 1000.00 \text{ ft}}{1315.21 \text{ ft} - 1000.00 \text{ ft}}\right)\\ &= N30°E\end{aligned}$$

$$\begin{aligned}\text{tangent bearing} \atop \substack{\text{out of curve} \\ \text{(PI to PT)}} &= \arctan\left(\frac{E}{N}\right)\\ &= \arctan\left(\frac{1524.01 \text{ ft} - 1181.99 \text{ ft}}{1439.69 \text{ ft} - 1315.21 \text{ ft}}\right)\\ &= N70°E\end{aligned}$$

$$\begin{aligned}\text{interior angle} &= \text{deflection angle at PI}\\ &= N70°E - N30°E\\ &= 40°\end{aligned}$$

$$\begin{aligned}\text{arc length} &= \frac{2\pi r(\text{interior angle})}{360°}\\ &= \frac{2\pi(1000 \text{ ft})(40°)}{360°}\\ &= 698.13 \text{ ft}\end{aligned}$$

Answer is (C)

104.

$$\text{correction} = \frac{(\text{applied tension} - \text{standard tension}) \times (\text{length})}{(\text{cross-sectional area}) \times (\text{modulus of elasticity})}$$

$$= \frac{(20 \text{ lb} - 10 \text{ lb})(100 \text{ ft})}{(0.003 \text{ in}^2)\left(30{,}000{,}000 \dfrac{\text{lb}}{\text{in}^2}\right)}$$

$$= \frac{1000 \text{ ft-lb}}{90{,}000 \text{ lb}}$$

$$= 0.011 \text{ ft}$$

$$\begin{aligned}\text{tape length} &= 100 \text{ ft} + 0.01 \text{ ft}\\ &= 100.01 \text{ ft}\end{aligned}$$

Answer is (B)

105.

$$\begin{aligned}\text{third angle} &= 180° - (90° + 30°)\\ &= 60°\\ A &= \tfrac{1}{2}ab\\ a &= 2\left(\frac{A}{b}\right)\\ &= 2\left(\frac{43{,}560 \text{ ft}^2}{b}\right)\\ &= \frac{87{,}120 \text{ ft}^2}{b}\end{aligned}$$

By the law of sines,

$$\frac{a}{\sin 30°} = \frac{b}{\sin 60°}$$

$$a = \frac{b\sin 30°}{\sin 60°}$$

$$= 0.57735\, b$$

$$\frac{87{,}120 \text{ ft}^2}{b} = 0.57735\, b$$

$$b^2 = \frac{87{,}120 \text{ ft}^2}{0.57735}$$

$$= 150{,}896.337 \text{ ft}^2$$

$$b = 388.45 \text{ ft}$$

Answer is (D)

106.

$$\text{angle BAC} = 180° - 115°02'$$
$$= 64°58'$$

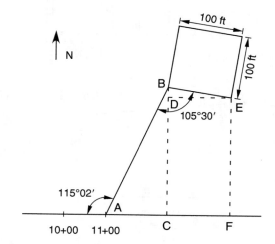

For triangle ABC,

$$\text{angle ABC} = 180° - (90° + 64°58')$$
$$= 25°02'$$
$$BC = AB \sin A$$
$$= 235.7 \text{ ft } (\sin 64°58')$$
$$= 213.56 \text{ ft}$$
$$AC = AB \sin B$$
$$= 235.7 \text{ ft } (\sin 25°02')$$
$$= 99.74 \text{ ft}$$

For triangle BDE,

$$\text{angle DBE} = 105°30' - 25°02'$$
$$= 80°28'$$
$$\text{angle DEB} = 180° - (90° + 80°28')$$
$$= 9°32'$$
$$BD = BE \sin E$$
$$= (100 \text{ ft})(\sin 9°32')$$
$$= 16.56 \text{ ft}$$
$$DE = BE \sin B$$
$$= (100 \text{ ft})(\sin 80°28')$$
$$= 98.62 \text{ ft}$$
$$\text{station} = 11{+}00 + AC + DE$$
$$= 1100 + 99.74 + 98.62$$
$$= 12 + 98.35$$
$$\text{offset} = BC - BD$$
$$= 213.56 \text{ ft} - 16.56 \text{ ft}$$
$$= 197.00 \text{ ft}$$

Answer is (C)

107.

$$\text{interior angle } (I) = \text{deflection angle at PI}$$
$$= \text{N65°E} - \text{S51°E}$$
$$= (90° - 65°) + (90° - 51°)$$
$$= 25° + 39°$$
$$= 64°$$
$$\text{arc length} = \left(\frac{I}{360°}\right) 2\pi r$$
$$= \left(\frac{64°}{360°}\right) 2\pi (2000 \text{ ft})$$
$$= 2234.02 \text{ ft}$$

Answer is (C)

108. Use double proportion measurement.

$$\Delta N \text{ (for N–S line)} = 12{,}805.35 - 4{,}847.92$$
$$= 7{,}957.43$$
$$\Delta E \text{ (for E–W line)} = 17{,}906.50 - 10{,}000.00$$
$$= 7{,}906.50$$

$$\begin{matrix}\text{north}\\\text{coordinate}\\\text{of the lost}\\\text{corner}\end{matrix} = \text{N (of S corner)}$$
$$+ \frac{(\text{record distance to south})(\Delta N)}{\text{record total north–south distance}}$$
$$= 4847.92 \text{ ft} + \frac{(80 \text{ ch})(7957.43)}{80 \text{ ch} + 40 \text{ ch}}$$
$$= 10{,}152.87 \text{ ft}$$

$$\begin{matrix}\text{east}\\\text{coordinate}\\\text{of the lost}\\\text{corner}\end{matrix} = \text{E (of W corner)}$$
$$+ \frac{(\text{record distance to west})(\Delta N)}{\text{record total east–west distance}}$$
$$= 10{,}000 + \frac{(79.25 \text{ ch})(7906.50)}{79.25 \text{ ch} + 40 \text{ ch}}$$
$$= 15{,}254.42 \text{ ft}$$

Answer is (C)

109.

	Monthly Mean Values (meters)		
	High Water	Low Water	Observed Range
Primary	5.65	4.05	1.60
New Station	3.29	1.89	1.40

By standard method,

$$\frac{\text{subordinate observed range}}{\text{subordinate mean range}} = \frac{\text{primary observed range}}{\text{primary mean range}}$$

$$\text{subordinate mean range} = \frac{(\text{primary mean range}) \times (\text{subordinate observed range})}{\text{primary observed range}}$$

$$= \frac{(1.45 \text{ m})(1.40 \text{ m})}{(1.60 \text{ m})}$$

$$= 1.27 \text{ m}$$

Answer is (A)

110. Curvature of the earth would increase the vertical distance from the line of sight down to the 95-foot elevation; refraction would counter the effect of curvature to some extent. The formula for the combined effect is

$$C = 0.667 \, M^2 - 0.093 \, M^2$$

$$= 0.574 \, M^2$$

$$= (0.574)(4.5 \text{ mi})^2$$

$$= 11.62 \text{ ft}$$

$$\text{line of sight} = 100.00 \text{ ft} + 11.62 \text{ ft}$$

$$= 111.62 \text{ ft}$$

$$\text{height above ground} = 111.62 \text{ ft} - 95.00 \text{ ft}$$

$$= 16.62 \text{ ft}$$

Answer is (C)

111. The collimation error would be proportional to the distance.

$$\frac{4.12 \text{ ft} - 4.00 \text{ ft}}{20 \text{ ft}} = \frac{x - 5.00 \text{ ft}}{180 \text{ ft}}$$

$$\frac{0.12 \text{ ft}}{20 \text{ ft}} = \frac{x - 5.00 \text{ ft}}{180 \text{ ft}}$$

$$x = 5.00 \text{ ft} + \frac{(0.12 \text{ ft})(180 \text{ ft})}{20 \text{ ft}}$$

$$= 6.08 \text{ ft}$$

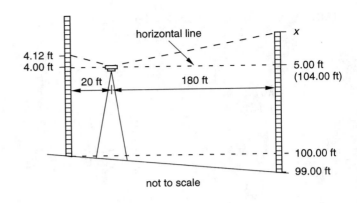

Answer is (D)

112. $\text{altitude angle of sun} = 90° - \text{zenith angle}$

$$= 90° - 55°56'06''$$

$$= 34°03'54''$$

If Z = azimuth to the sun,

$$\cos Z = \frac{\sin(\text{dec}) - \sin(\text{lat}) \sin(\text{alt})}{\cos(\text{lat}) \cos(\text{alt})}$$

$$= \frac{\sin 16°31'54'' - (\sin 38°10'06'')(\sin 34°03'54'')}{(\cos 38°10'06'')(\cos 34°03'54'')}$$

$$= \frac{(0.284545229) - (0.617973967)(0.560133056)}{(0.78619859)(0.828402655)}$$

$$= -0.09458538$$

Therefore,

$$Z = \arccos -0.09458538$$

$$= -95°25'39''$$

$$= 264°34'21''$$

$$\text{angle from target to sun} = 235°44'52''$$

$$\text{azimuth to target} = 28°49'29''$$

Answer is (A)

113. Δ height = cos (zenith angle)(slope distance)

$$= (\cos 85°)(450 \text{ ft})$$

$$= 39.22 \text{ ft}$$

$$\begin{array}{c}\text{ground}\\\text{elevation}\end{array} = 100.00 \text{ ft}$$

$$\text{HI} = 5.00 \text{ ft}$$

$$\begin{array}{c}\text{elevation}\\\text{of target}\end{array} = 144.22 \text{ ft}$$

Answer is (D)

114.

Angle (x)	$x - \bar{x}$	$(x - \bar{x})^2$
29.3″	2.7	7.29
24.0″	−2.6	6.76
27.9″	1.3	1.69
26.8″	0.2	0.04
26.1″	−0.5	0.25
25.9″	−0.7	0.49
26.1″	−0.5	0.25
27.8″	1.2	1.44
27.2″	0.6	0.36
28.0″	1.4	1.96
24.1″	−2.5	6.25
26.2″	−0.4	0.16
30.1″	3.5	12.25
29.7″	3.1	9.61
24.1″	−2.5	6.25
26.2″	−0.4	0.16
27.1″	0.5	0.25
24.9″	−1.7	2.89
25.7″	−0.9	0.81
25.2″	−1.4	1.96
$\overline{532.4″}$		$\overline{61.12}$

$$\text{mean} = \bar{x} = (31°02')\left(\frac{532.4″}{20}\right)$$

$$= 31°02'26.6″$$

$$\text{standard deviation} = \sqrt{\frac{\sum(x-\bar{x})^2}{n}}$$

$$= \sqrt{\frac{61.12″}{20}}$$

$$= 1.7″$$

Answer is (C)

115. Since there is 100 ft between the beginning of the vertical curve (PVC) and the PI,

$$\text{elevation PVC} = \text{elevation PI} - (\text{grade 1})(100 \text{ ft})$$

$$= 270.19 - (0.0125)(100 \text{ ft})$$

$$= 268.94 \text{ ft}$$

Answer is (C)

116. scale of negative = 1 in = 2000 ft

$$\text{scale of enlargement} = \frac{2000}{5}$$

$$= 400 \ (1 \text{ in} = 400 \text{ ft})$$

$$\begin{array}{c}\text{distance on}\\\text{enlargement}\end{array} = (24 \text{ in})\left(400 \frac{\text{ft}}{\text{in}}\right) = 9600 \text{ ft}$$

$$= (36 \text{ in})\left(400 \frac{\text{ft}}{\text{in}}\right) = 14{,}400 \text{ ft}$$

$$A = (9600 \text{ ft})(14{,}400 \text{ ft})$$

$$= 138{,}240{,}000 \text{ ft}^2 \left(\frac{1 \text{ ac}}{43{,}560 \text{ ft}^2}\right)$$

$$= 3174 \text{ ac}$$

Answer is (A)

117. When the end lot measurement is not given, all the excess or deficiency should be given to the end lot. Therefore, the monument for corner A should be set on line between the two block corners, at 200 ft from the northwest block corner.

$$\begin{array}{c}\text{bearing from}\\\text{NW to NE}\\\text{block corner}\end{array} = \arccos\left(\frac{\text{latitude}}{\text{departure}}\right)$$

$$= \arccos\left(\frac{10{,}006.89 - 10{,}000}{10{,}394.94 - 10{,}000}\right)$$

$$= \arccos 0.01744569$$

$$= \text{N89°00'01″E}$$

$$\begin{array}{c}\text{latitude from}\\\text{NW corner to A}\end{array} = (\text{distance}) \cos (\text{bearing})$$

$$= (200 \text{ ft})(\cos 89°00'01″)$$

$$= 3.49 \text{ ft}$$

$$\begin{array}{c}\text{departure from}\\\text{NW corner to A}\end{array} = (\text{distance}) \sin (\text{bearing})$$

$$= (200 \text{ ft})(\sin 89°00'01″)$$

$$= 199.97 \text{ ft}$$

$$\text{N coordinate} \atop \text{for A} = \text{N coordinate for NW corner}$$
$$+ \text{ latitude}$$
$$= 10{,}000 \text{ ft} + 3.49 \text{ ft}$$
$$= 10{,}003.49 \text{ ft}$$

$$\text{E coordinate} \atop \text{for A} = \text{E coordinate for NW corner}$$
$$+ \text{ departure}$$
$$= 10{,}000 \text{ ft} + 199.97 \text{ ft}$$
$$= 10{,}199.97 \text{ ft}$$

Answer is (D)

118. A deed call for one-half of a lot should be interpreted as a call for one-half of the area of the lot.

$$\text{total area of Lot 12} = \frac{(200 \text{ ft})(400 \text{ ft})}{2}$$
$$= 40{,}000 \text{ ft}^2$$

$$\text{area of the north one-half} = \frac{40{,}000 \text{ ft}^2}{2}$$
$$= 20{,}000 \text{ ft}^2$$

Answer is (D)

119. BS 5.24 ft BM A (elevation = 100.00 ft)

$$\text{HI} = 100.00 \text{ ft} + 5.24 \text{ ft} = 105.24 \text{ ft}$$

FS	Elev	Elev Above Base (91.14)	Offset
14.1	91.14	0	0
8.3	96.94	5.8	50
2.4	102.84	11.7	100
9.6	95.64	4.5	150
10.2	95.04	3.9	200
14.1	91.14	0	250

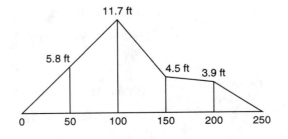

By the trapezoidal rule,

$$A = (50 \text{ ft}) \left(\frac{5.8 \text{ ft}}{2} \right)$$
$$+ (50 \text{ ft}) \left(\frac{5.8 \text{ ft}}{2} + 11.7 \text{ ft} + 4.5 \text{ ft} + \frac{3.9 \text{ ft}}{2} \right)$$
$$+ (50 \text{ ft}) \left(\frac{3.9 \text{ ft}}{2} \right)$$
$$= 145 \text{ ft}^2 + 1052.5 \text{ ft}^2 + 97.5 \text{ ft}^2$$
$$= 1295 \text{ ft}^2$$

Answer is (C)

120. Excess or deficiency between found, original monuments should be distributed among the intervening lots in proportion to their record measurements. Therefore, coordinates for corner A should be those of the NW corner plus four-fifths of the total difference in coordinates between the NW and NE corners of Block A.

$$\text{latitude} \atop \text{(NW to NE corners)} = 10{,}004.38 \text{ ft} - 10{,}000 \text{ ft}$$
$$= 4.38 \text{ ft}$$

$$\text{departure} \atop \text{(NW to NE corners)} = 10{,}250.96 \text{ ft} - 10{,}000 \text{ ft}$$
$$= 250.96 \text{ ft}$$

$$\text{latitude} \atop \text{(NW to A)} = \left(\frac{4}{5} \right) \text{latitude (NW to NE)}$$
$$= \left(\frac{4}{5} \right) (4.38 \text{ ft})$$
$$= 3.50 \text{ ft}$$

$$\text{departure} \atop \text{(NW to A)} = \left(\frac{4}{5} \right) \text{departure (NW to NE)}$$
$$= \left(\frac{4}{5} \right) (250.96 \text{ ft})$$
$$= 200.77 \text{ ft}$$

The coordinates for A are

$$N = N(NW) + \text{latitude (NW to A)}$$
$$= 10{,}000 \text{ ft} + 3.50 \text{ ft}$$
$$= 10{,}003.50 \text{ ft}$$
$$E = E(NW) + \text{departure (NW to A)}$$
$$= 10{,}000 \text{ ft} + 200.77 \text{ ft}$$
$$= 10{,}200.77 \text{ ft}$$

Answer is (C)

121. Under the same principle as in Question 120, corner B should be re-established halfway between the SW corner of Lot 9 and the SE corner of Lot 10.

$$\text{latitude} \atop \text{(SW to SE corners)} = 9804.38 \text{ ft} - 9802.64 \text{ ft}$$
$$= 1.74 \text{ ft}$$

$$\text{departure} \atop \text{(SW to SE corners)} = 10{,}252.45 \text{ ft} - 10{,}153.07 \text{ ft}$$
$$= 99.38 \text{ ft}$$

$$\text{latitude} \atop \text{(SW to B)} = \left(\frac{4}{5}\right) \text{latitude (SW to SE)}$$
$$= \tfrac{1}{2}(1.74 \text{ ft})$$
$$= 0.87 \text{ ft}$$

$$\text{departure} \atop \text{(SW to B)} = \left(\frac{4}{5}\right) \text{departure (SW to SE)}$$
$$= \tfrac{1}{2}(99.38 \text{ ft})$$
$$= 49.69 \text{ ft}$$

The coordinates for B are

$$N = N(SW) + \text{latitude (SW to B)}$$
$$= 9802.64 \text{ ft} + 0.87 \text{ ft}$$
$$= 9803.51 \text{ ft}$$
$$E = E(SW) + \text{departure (SW to B)}$$
$$= 10{,}153.07 \text{ ft} + 49.69 \text{ ft}$$
$$= 10{,}202.76 \text{ ft}$$

Answer is (D)

122. In the absence of evidence to the contrary, the exact width given on the plat should be used for streets. Therefore, corner C should be set 50 ft from the SE corner of Block A, on line between the SE corners of Blocks A and B.

$$\text{bearing from} \atop \text{Block A corner} = \arccos\left(\frac{\text{latitude}}{\text{departure}}\right)$$
$$= \arccos\left(\frac{9809.59 \text{ ft} - 9804.38 \text{ ft}}{10{,}551.41 \text{ ft} - 10{,}252.45 \text{ ft}}\right)$$
$$= \arccos 0.01744569$$
$$= N89°00'05''E$$

$$\text{latitude from block} \atop \text{corner to C} = (\text{distance}) \cos (\text{bearing})$$
$$= (50 \text{ ft})(\cos 89°00'05'')$$
$$= 0.87 \text{ ft}$$

$$\text{departure from block} \atop \text{corner to C} = (\text{distance}) \sin (\text{bearing})$$
$$= (50 \text{ ft})(\sin 89°00'05'')$$
$$= 49.99 \text{ ft}$$

$$\text{N coordinate for C} = \text{N coordinate for block corner} + \text{latitude}$$
$$= 9804.38 \text{ ft} + 0.87 \text{ ft}$$
$$= 9805.25 \text{ ft}$$

$$\text{E coordinate for C} = \text{E coordinate for block corner} + \text{departure}$$
$$= 10{,}252.45 \text{ ft} + 49.99 \text{ ft}$$
$$= 10{,}302.44 \text{ ft}$$

Answer is (D)

123. Parol evidence is oral testimony of witnesses.

Answer is (B)

124. The cite "112 So. 274" refers to the *Southern Regional Reporter*, Volume 112, page 274.

Answer is (B)

125. The Federal Emergency Management Administration (FEMA) publishes maps delineating flood-prone areas.

Answer is (D)

126. A design storm for calculating runoff quantities is a hypothetical rainfall representing the greatest amount of rain that statistically would be expected to fall over a specific period of time.

Answer is (D)

127. A square tract with an area of 1.5 ac would have sides

$$\sqrt{A} = \sqrt{(1.5 \text{ ac})\left(43{,}560 \; \frac{\text{ft}^2}{\text{ac}}\right)}$$
$$= \sqrt{65{,}340 \text{ ft}^2}$$
$$= 255.6 \text{ ft}$$

For a 22-in neat area,

$$\text{maximum scale} = \frac{\text{length of tract in feet}}{\text{length of sheet in inches}}$$
$$= \frac{255.6 \text{ ft}}{22 \text{ in}}$$
$$= 11.6 \; (1 \text{ in} = 11.6 \text{ ft})$$
$$\text{or } 1{:}11.6 \text{ in} \left(12 \; \frac{\text{ft}}{\text{in}}\right) = 1{:}139$$

The most desirable scale should be slightly less than this to allow for some margin. Therefore, the 1:200 scale would be best.

Answer is (B)

128. A period should be used to separate the two independent clauses.

Answer is (B)

129. At station 11+00,

$$\begin{array}{l}\text{difference} \\ \text{in elevation}\end{array} = (11{+}00 - 10{+}00)(-0.0042)$$

$$= -0.42 \text{ ft}$$

$$\begin{array}{l}\text{elevation of} \\ \text{sewer flow line}\end{array} = 105.50 \text{ ft} - 0.42 \text{ ft}$$

$$= 105.08 \text{ ft}$$

If the ground elevation = 115.60 ft, then

$$\text{cut} = 115.60 \text{ ft} - 105.08 \text{ ft}$$

$$= 10.52 \text{ ft}$$

Answer is (C)

130. Since the error (e) is proportional to the distance (x),

$$4.422 \text{ ft} - e = 4.467 \text{ ft} - 3e$$

$$2e = 4.467 \text{ ft} - 4.422 \text{ ft}$$

$$e = 0.0225 \text{ ft}$$

$$\text{correct reading} = 4.422 \text{ ft} - 0.022 \text{ ft}$$

$$= 4.400 \text{ ft}$$

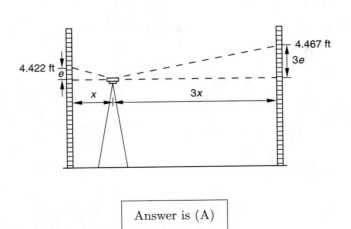

Answer is (A)

131. Dedication is the giving of land or rights in land to the public. Therefore, it is a means of transfer of rights to property.

Answer is (C)

132.

Field Notes

Station	+ HI	− Elevation	C/L
BM A		44.90	

3.30

11+00		4.4 7.1 4.9 5.2 5.1 9.1 11.2
		50 15 12 12 20 50

12+00		2.0 1.3 4.5 5.0 4.9 5.7 8.0
		50 18 12 12 18 50

BM B	5.15 43.05

From the field notes,

$$\text{HI} = 44.90 \text{ ft} + 3.30 \text{ ft} = 48.20 \text{ ft}$$

The ground elevation for any point on the cross sections would be HI − rod reading for that point. For section 12+00, right 18,

$$\text{ground elevation} = 48.20 \text{ ft} - 5.7 \text{ ft} = 42.5 \text{ ft}$$

Answer is (A)

133.

$$C = \frac{(105)(\sin B)}{2}$$

$$= \frac{(105)(\sin 45°)}{2}$$

$$= 37.12$$

Answer is (C)

134. The Soil Conservation Service of the U.S. Department of Agriculture (USDA) publishes soils maps for the U.S.

Answer is (A)

135. USGS 1:24,000 scale topographic maps cover 7.5′ of latitude and 7.5′ of longitude. One minute of latitude is equal to approximately one nautical mile or 6076 ft. Therefore, the north–south distance is

$$(7.5 \text{ min}) \left(6076 \ \frac{\text{ft}}{\text{min}}\right) = 45{,}570 \text{ ft}$$

Answer is (B)

136. A semicolon should be used to separate two sentences joined by a conjunctive adverb.

Answer is (A)

137. Generally, quarter-quarter corners are not established during public land field surveys. Government lots are also not laid out during original surveys. Therefore, both (A) and (B) are correct.

Answer is (D)

138. Of the four methods listed, the carrier phase method results in the most precise measurements. The carrier phase method "beats" or compares the satellite carrier phase with the signal from the local receiver oscillator.

Answer is (D)

139. The mean value is

$$\overline{X} = \frac{\sum \text{observations}}{\text{no. of observations}}$$

$$= \frac{\begin{array}{c}55'' + (60'' + 02'') + 59'' + 50'' + 58'' \\ + 59'' + 54'' + 56'' + (60'' + 00'')\end{array}}{9}$$

$$= 57''$$

Answer is (B)

140. To determine the median, sort the observations by value and then select the middle observation, which is 58″.

$50''$, $54''$, $55''$, $56''$, $\underline{58''}$, $59''$, $59''$, $(60'' + 00'')$, $(60'' + 02'')$

Answer is (C)

141. The mode is the most frequently encountered value and is therefore 59″.

Answer is (D)

142. The standard deviation is

$$\sigma = \sqrt{\frac{\sum\limits_{i=1}^{n} (Xi - \overline{X})^2}{n}}$$

$$= \frac{\sqrt{(2^2 + 5^2 + 2^2 + 7^2 + 1^2 + 2^2 + 3^2 + 1^2 + 3^2)}}{\sqrt{9}}$$

$$= 3.43''$$

Answer is (A)

143.

Target	D/R	Circle Reading	Angle (FS−BS)	Set Mean	Mean
1	D	000°00′34″			
3	D	157°41′41″	157°41′07″		
3	R	337°40′38″		157°41′06.5″	
1	R	179°59′32″	157°41′06″		
					157°41′04.5″
1	R	270°05′23″			
3	R	067°46′25″	157°41′02″		
3	D	247°47′24″		157°41′02.5″	
1	D	090°06′21″	157°41′03″		

The direct angle for the second position is 157°41′03″.

Answer is (A)

144. The difference between the means of angles for each set is

$$06.5'' - 02.5'' = 4''$$

Answer is (C)

145. Federal Geodetic Control Commission (FGCC) specifications require a rejection limit of 5″ of arc from the mean for all classes of both Second and Third Order work.

> Answer is (C)

146. For the binary number 110010,

$$
\begin{aligned}
\text{rightmost bit } 2^0 \times 0 &= 0 \\
2^1 \times 1 &= 2 \\
2^2 \times 0 &= 0 \\
2^3 \times 0 &= 0 \\
2^4 \times 1 &= 16 \\
2^5 \times 1 &= \underline{32} \\
&\ \ 50
\end{aligned}
$$

> Answer is (C)

147. Balancing sight lengths is the best method of minimizing error due to imperfect adjustment of a level.

> Answer is (D)

148.

magnetic bearing:	S85°15′W
magnetic azimuth:	265°15′
declination:	+8°30′ [add easterly declination]
true azimuth:	273°45′
true bearing:	N86°15′W

> Answer is (C)

149. Since a byte is comprised of 8 bits, the maximum decimal value it can store would be represented by an eight-bit binary number. An eight-bit binary can have a maximum value of 255(1+2+4+8+16+32+64+128).

> Answer is (C)

150. amount due = $50,000 + (0.16)$\left(\dfrac{6 \text{ months}}{12 \text{ months}} \right)$

$$\times \ \$50,000$$

$$= \$54,000$$

> Answer is (C)

151. Under the early Spanish system of measurement, a league was a unit of both linear and area measurement. A league is 5000 varas or approximately 13,889 ft in length. It is also a square 5000 varas on a side, or approximately 4428 ac.

> Answer is (C)

152.

direct labor:		$4520
overhead:	(150%)($4250) =	$6780
		$11,300
operating margin:	(12%)($11,300) =	$1356
total fee:		$12,656

> Answer is (D)

153. In the description provided, the westerly right-of-way of Highway 20 is an example of a bounds, which is a monument or adjoiner called for in the deed.

> Answer is (C)

154. Other than by use of a COGO routine, the most expedient method of determining area of the described tract is by dividing it into two triangles and determining the area of each by the following formula:

$$A = \tfrac{1}{2}ab \sin C$$

a and b are two sides of the triangle, and C is the included angle.

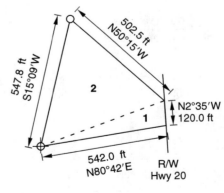

For triangle 1,

$$A = \tfrac{1}{2}(542 \text{ ft})(120 \text{ ft})(\sin 96°43')$$
$$= 32,296.80 \text{ ft}^2$$

For triangle 2,

$$A = \tfrac{1}{2}(502.5 \text{ ft})(547.81 \text{ ft})(\sin 65°23'42'')$$
$$= 125{,}139.77 \text{ ft}^2$$
$$\text{total area} = 32{,}296.80 \text{ ft}^2 + 125{,}139.77 \text{ ft}^2$$
$$= 157{,}436.57 \text{ ft}^2$$
$$\frac{157{,}436.57 \text{ ft}^2}{43{,}560 \frac{\text{ft}^2}{\text{ac}}} = 3.61 \text{ ac}$$

Answer is (C)

155. allowable misclosure in mm $= (8)\left(\sqrt{D}\right)$

$$D = \text{distance in km}$$

$$\text{misclosure} = (8)\left[\sqrt{(1 \text{ mi})\left(1.6 \frac{\text{km}}{\text{mi}}\right)}\right]$$
$$= 10.1 \text{ mm}$$

Answer is (B)

156. Hachures are a method of showing relief on maps. The symbol consists of short, nearly parallel lines drawn parallel to the direction of slope.

Answer is (C)

157. ALTA/ACSM minimum requirements for urban (Class A) surveys require a linear closure, after angle balancing, of 1:15,000. For a length of 3000 m, the closure is

$$\frac{3000 \text{ m}}{15{,}000} = 0.2 \text{ m}$$

Answer is (B)

158. For Section 16, the quarter corners on the east, south, and west sides should be re-established by single proportional measurement. Since all three lines were 80 chains in length with quarter corners set at 40 chains, the quarter corners should be re-established at the midpoints of the three lines. The center of the section can then be established at the intersection of the two lines connecting the four quarter corners.

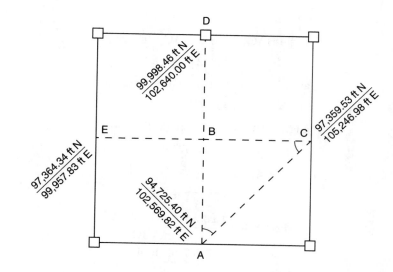

east quarter: $N = \dfrac{99{,}996.93 + 94{,}722.13}{2}$
$$= 97{,}359.53 \text{ ft}$$
$$E = \dfrac{105{,}270.00 + 105{,}223.97}{2}$$
$$= 105{,}246.98 \text{ ft}$$

south quarter: $N = \dfrac{94{,}722.13 + 94{,}728.67}{2}$
$$= 94{,}725.40 \text{ ft}$$
$$E = \dfrac{105{,}223.97 + 99{,}915.66}{2}$$
$$= 102{,}569.82 \text{ ft}$$

west quarter: $N = \dfrac{100{,}000 + 94{,}728.67}{2}$
$$= 97{,}364.34 \text{ ft}$$
$$E = \dfrac{100{,}000 + 99{,}915.66}{2}$$
$$= 99{,}957.83 \text{ ft}$$

section center:

bearing $AC = \arctan\left(\dfrac{Ec - Ea}{Nc - Na}\right)$
$$= \text{N}45°27'51''\text{E}$$

bearing $AB = \arctan\left(\dfrac{Ea - Ed}{Na - Nd}\right)$
$$= \text{N}0°45'51''\text{E}$$

bearing $CB = \arctan\left(\dfrac{Ec - Ee}{Nc - Ne}\right)$
$$= \text{N}89°56'52''\text{W}$$

$$\text{angle BAC} = \text{bearing AC} - \text{bearing AB}$$
$$= 44°42'06''$$
$$\text{angle ACB} = \text{bearing CB} - \text{bearing CA}$$
$$= 44°35'17''$$
$$\text{angle ABC} = 180° - \text{angle BAC} - \text{angle ACB}$$
$$= 90°42'37''$$
$$\text{distance AC} = \sqrt{(Ec-Ea)^2 + (Nc-Na)^2}$$
$$= 3755.77 \text{ ft}$$
$$\text{distance AB} = \frac{(\text{distance AC})(\sin \text{ACB})}{\sin \text{ABC}}$$
$$= \frac{(3755.77)(\sin 44°35'17'')}{\sin 90°42'37''}$$
$$= 2636.77 \text{ ft}$$
$$\text{latitude AB} = \text{distance AB} + \cos(\text{bearing AB})$$
$$= 2636.77 + \cos 0°45'51''$$
$$= 2636.54 \text{ ft}$$
$$Nb = Na + \text{latitude AB}$$
$$= 94{,}725.40 \text{ ft} + 2636.54 \text{ ft}$$
$$= 97{,}361.94 \text{ ft}$$
$$\text{departure AB} = (\text{distance AB}) \sin(\text{bearing AB})$$
$$= (2636.77 \text{ ft})(\sin 0°45'51'')$$
$$= 35.09 \text{ ft}$$
$$Eb = Ea + \text{departure AB}$$
$$= 102{,}569.82 \text{ ft} + 35.09 \text{ ft}$$
$$= 102{,}604.91 \text{ ft}$$

Answer is (D)

159. Scanning typically creates a grid or raster structure file.

Answer is (C)

160. The stationing of Point A can be determined by solving the triangle shown.

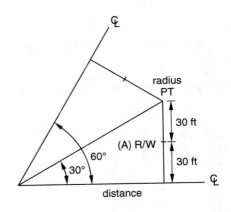

$$\tan 30° = \frac{60 \text{ ft}}{\text{distance}}$$
$$\text{distance} = \frac{60 \text{ ft}}{\tan 30°}$$
$$= 103.92 \text{ ft}$$
$$\text{stationing} = 12+65$$
$$\text{plus distance} \quad \underline{1+03.92}$$
$$13+68.92$$

Answer is (B)

161. The distance from the intersection of the two centerlines to Point A is the hypotenuse of the illustrated right triangle.

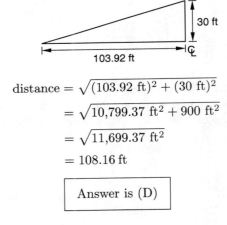

$$\text{distance} = \sqrt{(103.92 \text{ ft})^2 + (30 \text{ ft})^2}$$
$$= \sqrt{10{,}799.37 \text{ ft}^2 + 900 \text{ ft}^2}$$
$$= \sqrt{11{,}699.37 \text{ ft}^2}$$
$$= 108.16 \text{ ft}$$

Answer is (D)

162.
$$\tan(90° - 75°25') = \frac{\text{vertical distance}}{1575 \text{ m}}$$
$$\text{vertical distance} = (\tan 14°35')(1575 \text{ m})$$
$$= 409.77 \text{ m}$$

$$409.77 \text{ m}$$
$$+ \; 52.60 \text{ m [ground elevation]}$$
$$+ \quad 4.5 \text{ m [height of instrument]}$$
$$\overline{466.87 \text{ m}}$$

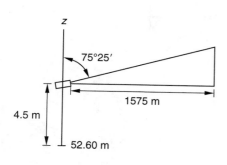

Answer is (D)

165.

$$Ls = 1.6\frac{V^3}{R}$$

$$V = \sqrt[3]{\frac{(Ls)(R)}{1.6}}$$

$$= \sqrt[3]{\frac{(300 \text{ ft})(1200 \text{ ft})}{1.6}}$$

$$= \sqrt[3]{225{,}000}$$

$$= 60.8 \text{ mph}$$

Answer is (C)

163.

$$\Delta 2 = \Delta \text{total} - \Delta 1$$

$$= 68°00' - 35°00'$$

$$= 33°00'$$

$$L_1 = \left(\frac{\Delta 1}{360°}\right) 2\pi r_1$$

$$= \left(\frac{35°}{360°}\right) 2\pi (600 \text{ ft})$$

$$= 366.52 \text{ ft}$$

$$L_2 = \left(\frac{\Delta 2}{360°}\right) 2\pi r_2$$

$$= \left(\frac{33°}{360°}\right) 2\pi (400 \text{ ft})$$

$$= 230.38 \text{ ft}$$

$$\text{PT station} = \text{PC} + L_1 + L_2$$

$$= (10{+}46.32) + 366.52 \text{ ft} + 230.38 \text{ ft}$$

$$= 16{+}43.22 \text{ ft}$$

Answer is (C)

166.

Line	Bearing	Length (ft)	Latitude $L \cos$ (bearing)	Departure $L \sin$ (bearing)
AB	S67°45'E	449.83	−170.33	+416.34
BC	N28°19'E	560.27	+493.23	+265.76
CD	N56°23'W	484.18	+268.06	−403.21
DE	S08°50'W	375.42	−370.97	−57.65
EA			(−219.99)	(−221.24)
			0.00	0.00

$$\text{length of EA} = \sqrt{(219.99 \text{ ft})^2 + (221.24 \text{ ft})^2}$$

$$= \sqrt{48{,}395.60 \text{ ft}^2 + 48{,}947.14 \text{ ft}^2}$$

$$= \sqrt{97{,}342.74 \text{ ft}^2}$$

$$= 312.00 \text{ ft}$$

Answer is (D)

164.

$$\text{deflection angle} = \frac{\Delta 1}{2} = \frac{35°}{2}$$

$$= 17°30'$$

Answer is (B)

167. The total section area is 640 ac.

$$\text{one quarter} = \frac{640 \text{ ac}}{4} = 160 \text{ ac}$$

$$\text{one quarter-quarter} = \frac{160 \text{ ac}}{4} = 40 \text{ ac}$$

$$\text{one quarter-quarter-quarter} = \frac{40 \text{ ac}}{4} = 10 \text{ ac}$$

$$\text{one half of one quarter-quarter-quarter} = \frac{10 \text{ ac}}{2} = 5 \text{ ac}$$

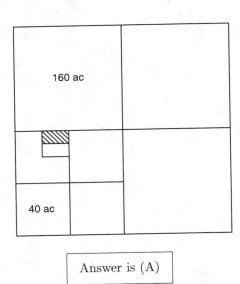

Answer is (A)

168. From the illustration, the described tract lies westerly of the $N\frac{1}{2}$ of the $NE\frac{1}{4}$ of the $NE\frac{1}{4}$ of the $SW\frac{1}{4}$.

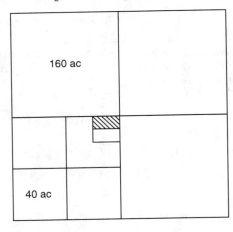

Answer is (D)

169.

$$\tan \phi = \frac{1}{10,000}$$

$$\phi = \arctan\left(\frac{1}{10,000}\right)$$

$$= 20.6''$$

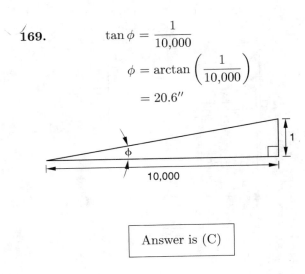

Answer is (C)

170. angular uncertainty

$$\text{for } 20'' = (\tan 20'')(1500 \text{ ft})$$

$$= 0.15 \text{ ft}$$

linear uncertainty

$$\text{for } 1{:}10,000 = \frac{1500 \text{ ft}}{10,000}$$

$$= 0.15 \text{ ft}$$

$$0.15 \text{ ft} + 0.15 \text{ ft} = 0.30 \text{ ft}$$

Answer is (D)